Annals of
KIRSTENBOSCH
Botanic Gardens

Volume 17

J N ELOFF DSc
Editor
Executive Director, National Botanic Gardens

Lachenalia aloides (L.f.) Engl. var. *quadricolor* (Jacq.) Engl. (reproduced from an original watercolour drawing by Miss W.F. Barker.)

THE LACHENALIA HANDBOOK

A guide to the genus, with introductory notes on history, identification and cultivation, with descriptions of the species and colour illustrations

GRAHAM DUNCAN

Kirstenbosch National Botanic Gardens

National Botanic Gardens
1988

© National Botanic Gardens

ISSN 0 258-3305

ISBN 0 620 11953 5

Colour reproduction by Sparhams, Cape
Printed by CTP Book Printers, Cape

BD8102

CONTENTS

FOREWORD

It is indeed a pleasure to introduce The Lachenalia Handbook, an addition to the Annals of Kirstenbosch Botanic Gardens. The standard of this Handbook is such that it will grace this collection of publications and at the same time its inclusion in this series gives it, at the outset, a stamp of approval which is richly deserved. It is an attractive, informative, accurate and practical handbook and it amply satisfies the educational goal of the National Botanic Gardens, as defined and pursued by the Board and the staff:

"To promote a greater appreciation of our indigenous flora by conveying knowledge of plants and the environment to as wide a public as possible." One way of achieving this is to "publish information and results of observations and research of staff and other authors". This book is the result of four years of horticultural research on the part of the author, Graham Duncan.

On a less formal note, it also gives me great personal pleasure to welcome this book. To the many people who admire and enjoy lachenalias the availability of a well-written, attractively illustrated handbook which provides simple but adequate horticultural advice is a dream come true. Its purpose may have to stretch even a little further. While the botanical world continues to wait for the definitive treatise of the genus, on which Miss W. F. Barker has been engaged for many years, Graham Duncan's publication will meet an interim need, in addition to the popular appeal it will make to flower lovers of all ranks.

Miss Barker has contributed an historical background; this, added to the very up-to-date inclusion of the species recently described and named by Miss Barker and by others covers almost the whole field as it is known at present. The rate of recent activity in *Lachenalia* taxonomy is indicated by the inclusion in the handbook of 37 new species published during the last 10 years. During the same period quite a number of familiar names were dropped. When Kirstenbosch Gardens accepted my personal collection of living lachenalias in 1977, it contained 61 named specimens out of about 72 then published or putative species; the Handbook now lists a total of 88 species.

What a delight it is to recognize several then anonymous favourites now respectably named!

The author is to be congratulated for the quality of this work and he must be thanked for the joy this book will bring to many people; these compliments are also extended to the publishers, the National Botanic Gardens.

M. C. BOTHA
Chairman of the Board: March 1988
National Botanic Gardens

LACHENALIA Jacq.f. ex Murrary

L. algoensis Schonl.
L. aloides (L.f.) Engl.—(=*L. tricolor* Jacq.f.)
L. ameliae W.F. Barker
L. angelica W.F. Barker
L. anguinea Sweet
L. arbuthnotiae W.F. Barker
L. bachmannii Bak.
L. barkeriana U. Müller-Doblies *et al.*
L. bolusii W.F. Barker
L. bowkeri Bak.
L. buchubergensis Dinter
L. bulbifera (Cyrillo) Engl.— (=*L. pendula* Ait.)
L. capensis W.F. Barker
L. carnosa Bak.—(=*L. ovatifolia* L. Guthrie)
L. comptonii W.F. Barker
L. concordiana Schltr. ex W.F. Barker
L. congesta W.F. Barker
L. contaminata Ait.
L. dasybotrya Diels
L. dehoopensis W.F. Barker
L. duncanii W.F. Barker
L. elegans W.F. Barker
L. esterhuysenae W.F. Barker
L. fistulosa Bak.—(=*L. convallariodora* Stapf)
L. framesii W.F. Barker
L. giessii W.F. Barker
L. gillettii W.F. Barker
L. glaucophylla W.F. Barker
L. haarlemensis Fourc.
L. hirta (Thunb.) Thunb.
L. isopetala Jacq.
L. juncifolia Bak.
L. klinghardtiana Dinter
L. kliprandensis W.F. Barker
L. latifolia Tratt.
L. latimerae W.F. Barker
L. leomontana W.F. Barker
L. liliflora Jacq.
L. longibracteata Phillips
L. macgregoriorum W.F. Barker
L. margaretae W.F. Barker
L. marginata W.F. Barker
L. martinae W.F. Barker
L. mathewsii W.F. Barker
L. maximiliani Schltr. ex W.F. Barker

L. mediana Jacq.
L. minima W.F. Barker
L. moniliformis W.F. Barker
L. montana Schltr. ex W.F. Barker
L. muirii W.F. Barker
L. multifolia W.F. Barker
L. mutabilis Sweet
L. namaquensis Schltr. ex W.F. Barker
L. namibiensis W.F. Barker
L. nordenstamii W.F. Barker
L. orchioides (L.) Ait.—(=*L. glaucina* Jacq.)
L. orthopetala Jacq.
L. pallida Ait.
L. patula Jacq.—(=*L. succulenta* Masson ex Bak.)
L. pearsonii (Glover) W.F. Barker—(=*Scilla pearsonii* Glover)
L. peersii Marloth ex W.F. Barker
L. physocaulos W.F. Barker
L. polyphylla Bak.
L. polypodantha Schltr. ex W.F. Barker
L. purpureo-caerulea Jacq.
L. pusilla Jacq.—(=*Polyxena pusilla* (Jacq.) Schltr.)
L. pustulata Jacq.
L. reflexa Thunb.
L. rosea Andrews
L. rubida Jacq.
L. salteri W.F. Barker
L. sargeantii W.F. Barker
L. schelpei W.F. Barker
L. splendida Diels—(=*L. roodeae* Phillips)
L. stayneri W.F. Barker
L. trichophylla Bak.—(=*L. massonii* Bak.)
L. undulata Masson ex Bak.
L. unicolor Jacq.
L. unifolia Jacq.
L. variegata W.F. Barker
L. ventricosa Schltr. ex W.F. Barker
L. verticillata W.F. Barker
L. violacea Jacq.
L. viridiflora W.F. Barker
L. whitehillensis W.F. Barker
L. youngii Bak.
L. zebrina W.F. Barker
L. zeyheri Bak.

ALPHABETICAL LIST OF COLOUR PLATES

All photos are by the author except for those acknowledged.

ACKNOWLEDGEMENTS

This publication is dedicated to Miss W. F. Barker, without whose expertise it could not have been attempted. Her studies over many years have contributed so greatly to our knowledge of *Lachenalia*, and laid a solid foundation for further research. In addition to checking the manuscript, Miss Barker also kindly contributed the notes on the historical background to the genus.

I am very grateful indeed to the many other people who helped with this work. In particular I would like to thank Mr John Winter for his co-operation throughout the four-year period of this project, Dr John Rourke for advice on numerous matters and Deirdré Snijman for her helpful comments on the manuscript. My colleagues have enthusiastically assisted me in collecting material in the wild, and I am particularly thankful to Mrs Margaret Thomas, Pauline Perry, Ernst van Jaarsveld and Dr Kim Steiner. A special word of thanks is due to Hendrik and Rhoda van Zijl for their assistance in many ways over the years.

The black and white illustration was prepared by Mrs Ellaphie Ward-Hilhorst, and the following people kindly lent me slides for publication: Miss W. F. Barker; the late Mr Percy Sargeant; the late Mr Harry Hall; Mrs Ann Scott; Prof. M. C. Botha, the late Rev. T. M. Wurts and Prof. D. Müller-Doblies.

I am also very grateful to Desirée du Bois and Sonja Steinhobel for typing the manuscript and to Lovell Bosman who sorted out all the editorial details.

Very special thanks go to my mother, Judy, with whom I have shared an interest in lachenalias for many years.

Finally my thanks go to the Executive Director and the Chairman of The Board of National Botanic Gardens, under whose authority this work was undertaken.

INTRODUCTION

In the absence of a definitive monograph of the genus, this "Lachenalia Handbook" is an attempt to collate available information so as to provide horticulturists and informed gardeners with a list of valid species names with synonyms, illustrations of as many species as possible and notes on identification, cultivation and other relevant topics.

A formal key to the genus has not been included in this publication; it is currently in preparation and will be published at a later date by Miss W. F. Barker. The species descriptions have however been arranged in a manner which, taken in conjunction with the illustrations, will assist the user in providing some aid to identification.

This Handbook is based on the extensive dried plant collection housed in the Compton Herbarium at Kirstenbosch, largely assembled by Miss W. F. Barker over many years and the large living collection maintained by the author in the Kirstenbosch bulb nursery. Observation of species in their natural habitat has also played an important role in this publication.

Although the differences between most species are sufficiently distinct to make them easily recognized, there are a number of "complex" species consisting of many forms which grade into one another, and are difficult to separate into distinct species or varieties. As an example, we may take *L. aloides* which, although having certain clear-cut varieties, also has a multitude of other forms which tend to show slight variation from locality to locality, and sometimes even obvious variation within a particular locality.

Various chromosome studies on the genus have been undertaken in the past (Moffett, 1936; de Wet, 1957); in a recent survey (Ornduff and Watters, 1978), forty-one collections consisting of sixteen named species and various others of uncertain identity, a remarkable variability in chromosome number was found with a minimum value of 2n = 10 rising to 2n = 56. Present comprehensive cytological studies being carried out in the United Kingdom (Brandham, unpublished), may yield further important results which will no doubt be of great value in unravelling "complex" species and in determining accurate species limits.

With new *Lachenalia* species continually coming to light and the immense variability which seems to exist within many species, it is clear that the genus is not as yet taxonomically sufficiently understood and that much research has still to be done.

The author does not claim the information contained here to be absolutely comprehensive, particularly with regard to distribution ranges given, nor that the species arrangement is without shortcomings. Observations of lachenalias in the wild from readers will be greatly appreciated.

HISTORICAL BACKGROUND (BY MISS W. F. BARKER)

The genus *Lachenalia* has had a long, interesting, eventful but complicated historical background, extending over just more than three centuries, from

1685/6–1987. A vast amount of taxonomic literature has been published on the genus, from which only some of the most important highlights can be dealt with here. The three hundred years can be conveniently divided into five periods.

I. 1685/6–1691: THE PERIOD OF THE VERY EARLY CODICES OR FLORILEGIA

After the establishment of the Dutch East India Company's Garden at the Cape in 1652, explorations into the interior began. This resulted in many new introductions into European gardens. Interest in Cape plants accelerated and they became fashionable. The scientists of the day began to make collections of water-colour paintings of those new and curious plants; these were known as Codices or Florilegia. Many paintings were copied or recopied for favoured patrons, and it was not unusual for paintings of the same plant to appear in several different codices. Eventually they developed into important scientific documents composed of hundreds of paintings and became highly prized possessions and family heirlooms; some of them were bound and embellished with family crests. Fortunately, a number of them have survived after having passed through the hands of several distinguished owners. Several have been described and published this century and their history makes fascinating reading.

Seven of the Codices are known to include *Lachenalia* paintings and four include paintings of the same plant now known as *Lachenalia hirta* (Thunb.) Thunb. The most important of these is in Trinity College Library, Dublin (TDC). It was described by Gilbert Waterhouse in 1932. In it he stated: "The Trinity College Ms is the identical section removed about 1691–1692 from the Archives of the Dutch East India Company, and is considered to be the one Simon van der Stel used to illustrate the diary of his expendition to Namaqualand in 1685/6." The painting of *Lachenalia hirta* in it can therefore be taken to be the earliest record of the genus in colour to which a definite date can be ascribed.

Four of the other Florilegia deal mainly with the paintings of the plants now known as *Lachenalia orchioides* (L.) Ait., *Lachenalia glaucina* Jacq. and *Lachenalia contaminata* Ait. The most important of these is "The Flora Capensis of Jakob and Johann Phillip Breyne" (BFC), described by Mary Gunn and Enid du Plessis in 1978. The volume is housed in the Brenthurst Library in Johannesburg.

Under the heading "Indication of the Ownership of the Volume" the authors say: "The fact that the Breyne coat of arms appears on the front cover indicates that the volume was bound for the Breyne Library, for Johann Phillip Breyne whose coloured book plate featuring the family crest is pasted onto the verso of the first leaf." Two more family crests indicate subsequent ownership and in 1956 the volume was sold to Sir Ernest Oppenheimer, and the family bookplate has been added.

II. 1692–1752: THE PERIOD OF THE PRELINNAEN PUBLICATIONS

1692: The distinguished scientific authors of the Prelinnaen Publications relied to a great extent on the early codices for their information, and the familiar figure of *Lachenalia hirta* was the first to appear in print in "Leonardi Plukenetii Phytographia" on tab. CXCV Fig. 5. He described it as "Hyacinthus Africanus Orchioides serpentarius, folio singularis, undato. piliscilliaribus fimbriato, floribus ex aureo punicatibus" and added, "Codicis Comptoniani", indicating that he had taken it from the codex of Bishop Compton of London; the whereabouts of this codex appears to be unknown at present.

1709: James Petiver, Apothecary, published his "Gazophylacii Naturae & Artis, Herbarium Capense" in 1709. The illustration in it is an exact mirror image of Plukenet's figure of *Lachenalia hirta*. He states that the figures were "Copied from the original paintings taken from the living plants, viz. those which the States of Amsterdam presented to the Right Reverend the Bishop of London, when his Lordship was at the Congress there A.D. 1691."

1739: The most important Prelinnaen Publication in this period is "Jacobi Breynii Plantarum et Icones Rariorum et Exoticarum Plantarum". In it the three lachenalias from the Codex "Flora Capensis of Jakob & Johann Phillip Breyne" (BFC) now in the Brenthurst Library are all depicted on the same engraved plate. Fig. 1 described as "Hyacinthus orchioides, Africanus maior bifolius flore caerulea" represents *Lachenalia glaucina* Jacq. Fig. 2 "Hyacinthus orchioides, Africanus maior, bifolius maculatus, flore sulphureo majore" represents *Lachenalia orchioides* (L.) Ait. and Fig. 3. "Hyacinthus orchioides, aphyllus, serpentarius maior" represents *Lachenalia contaminata* Ait., which has many filiform leaves.

III. 1753–1784: The Period Between the Publication of Linnaeus's Species Plantarum in 1753 and J. A. Murray's Inadvertant Publication of the Genus Lachenalia.

,1753: When Linnaeus introduced his binomial system in "Linnaeus Species Plantarum" Ed. I: 318 (1783), he described the first taxon which is now included in the genus *Lachenalia:*

HYACINTHUS
Orchioides 11 HYACINTHUS corollis irregularibus sexpartitis
      Hyacinthus orchioides africanua bifolius maculatus, flore sulphureo
      obsoleto majore. Breyne, prod. 3.p.24.t.11.f.2. Habitat in Aethio-
      pia.

Thus citing Fig. 2 as the iconotype of *Hyacinthus orchioides* L. which matches the painting in the Codex "The Flora Capensis" of Jakob and Johann Phillip Breyne (BFC) in the Brenthurst Library. It later became *Lachenalia orchioides* (L.) Ait.

1781: In "Supplementarum Plantarum et Specierum Plantarum" Ed. II, Linnaeus fil. published *Phormium aloides* L.f. later to become *Lachenalia aloides* (L.f.) Engler.

1784: Carl Peter Thunberg who came to be known as the "Father of South African Botany" and famous for his three journeys of Botanical exploration into the interior during 1772–1775, published his "Dissertio de Novis Generibus Plantarum", which included the genus *Phormium*. He included four species, *Phormium tenax* "The New Zealand flax", *Phormium aloides* and *Phormium orchioides* with five varieties, four of which are now recognized as good species, this illustrates the difficulties of interpreting the species in this large and bewildering genus. His fourth species *Phormium hirtum* was the plant which had been illustrated in the Codex of Simon van der Stel's Journal of his expedition to Namaqualand in 1685/6 a century earlier.

1784: In the meantime collections in the Schönbrunn Palace Gardens near Vienna were increasing rapidly, and were being studied by Baron Nicolaus Joseph Jacquin and his son Joseph Franz Jacquin. The latter had described a new genus naming it *Lachenalia* after Werner de Lachenal, an eminent professor in Basel, Switzerland, and giving it the specific name *tricolor* for its three-coloured flower. In 1780 he sent the manuscript to the editor of the journal "Acta Helvetica", in Basel, expecting it to be published in that year. However, the publication of the journal had lapsed and his paper was ultimately only published in the revived journal titled "Nova Acta Helvetica", in 1787. J. A. Murray, evidently having seen the manuscript and presuming that it had been published, included a short description of the genus in "Linnaeus Systema Vegetabilium" Ed. 14 1784, citing the specific name as *tricolor*. Although Murray had not intended to publish the new genus, he must be credited with having done so, and the correct citation for it is *Lachenalia* Jacq.f. ex Murray. However the species had earlier been published as *Phormium aloides* L.f. and the correct citation therefore now is *Lachenalia aloides* (L.f.) Engler.

IV. 1786–1897: The Period Between N. J. Jacquin's "Icones Plantarum Rariorum" and J. G. Baker's Final Monograph with a Key to the Species.

This was a period during which many taxonomic works were published, and chief among them were N. J. Jacquin's magnificent illustrated volumes. In Vol. I of his "Icones Plantarum Rariorum" dated 1781–1786 on the title page, the colour plate of *Lachenalia tricolor* Jacq.f., originally intended to appear in "Acta Helvetica" Vol. 9, was published, and in Vol. II, twenty-two colour plates of *Lachenalia* species described in his "Collectanea", were published, ten of which are iconotypes of his species, the holotypes of which were destroyed during the bombing of Vienna in World War II.

1786: The first volume of the world-renowned illustrated publication "Curtis's Botanical Magazine" was published in 1786, but it was not until 1789 that a *Lachenalia* plate appeared in Vol. 3 on t.82. It was named *Lachenalia tricolor*; however, it was not the plant which Jacquin f. had in mind when he described his species, but another variety of *Lachenalia aloides* (L.f.) Engler.

Lachenalia ameliae
(Montagu)

1b. *Lachenalia ameliae*
(Karoo Poort)

1c. *Lachenalia arbuthnotiae*
(Wynberg)

1d. *Lachenalia carnosa*
(Kamieskroon)

2a. *Lachenalia carnosa*
(Nababeep)

2b. *Lachenalia carnosa*
(Nababeep)

2c. *Lachenalia concordiana*
(Calvinia)

2d. *Lachenalia co*
(Suther

1789: William Aiton's "Catalogue of the species cultivated in the Botanic Garden Kew" was published in 1789. He adopted the new generic name for the genus and included six *Lachenalia* species.

1794: Thunberg revised his work on the genus in his "Prodromus Plantarum Capensium" in 1794, under the generic name *Lachenalia*, raising one of his earlier varieties to specific rank as *Lachenalia reflexa* Thunb., and he made the new combination of *Phormium hirtum* Thunb. to *Lachenalia hirta* (Thunb.) Thunb.

1871: Many new editions of previous works followed, dealing mainly with the species known to Jacquin, Thunberg and Aiton, until in 1871 J. G. Baker produced his first monograph in "The Journal of the Linnaen Society" (Botany) Vol. XI. He included twenty-nine species, provided an indented Key, and grouped related species into subspecies for the first time.

Subsequently as new taxa became available they were described and published by various authors, either singly or in groups, in many journals.

1891–1898: Rudolph Schlechter on his collecting trips in Southern Africa, made a large and valuable collection of *Lachenalia* species, many of them undescribed. He gave many of them his manuscript names but none was published until well into the twentieth century.

1896–1897: J. G. Baker published his major and final work on the genus in "The Flora Capensis" Vol. VI; in it the number of species had increased to forty-two. He cited many herbarium specimens with their precise localities where possible, their collectors and their collecting numbers, which made reference to them more accessible.

V. 1900–1987: THE PERIOD OF THE TWENTIETH CENTURY

1921: The magnificent illustrated publication "The Flowering Plants of South Africa" which is to South Africa what "Curtis's Botanical Magazine" is to Great Britain, was launched in 1921. The earliest *Lachenalia* plate to be published in it was *Lachenalia roodeae* Phill. in Vol. 3 Pl. 91 (1923); since then sixteen more have been added.

No major work on the genus *Lachenalia* has been published during the twentieth century, but the foundation for it has been laid at the National Botanic Gardens, Kirstenbosch, which was founded in 1913, and where most of the taxonomic work has been done.

1929: By 1929 a working collection of living plants had been assembled in the Nursery, which was studied and illustrated at the Bolus Herbarium, which was then housed in the grounds at Kirstenbosch.

1935: The new taxonomic "Journal of South African Botany", with Professor R. H. Compton, the Director, as its editor, made its first appearance in 1935. It served as the main vehicle for the publication of the new *Lachenalia* species until 1984, when it was amalgamated with "The South African Journal of Botany", in which new species now continue to be published.

1940: The nucleus of the new Kirstenbosch Herbarium was transferred from the Director's office to the building vacated in 1937 by the Bolus Herbarium. It now houses a large collection of herbarium specimens of *Lachenalia* species, many of which are holotypes; also paintings and colour photographs.

1961: The "Indigenous Bulb Growers Association of South Africa" (IBSA) was established on 15th April 1961. It now has a world-wide membership. A questionnaire among all its members in 1985 revealed that the genus *Lachenalia* came second only to *Gladiolus* in popularity.

1987: The living collection of *Lachenalia* species in the Kirstenbosch Nursery has expanded over the years largely due to the efforts of Graham D. Duncan, in charge of the bulb section, and to Margaret L. Thomas who was instrumental in establishing IBSA. With new species being added so frequently, the preoccupation has been to describe, and latterly to photograph and name as many as possible, for as Jacquin is reputed to have said "A genus is not known until all its species are perfectly understood". The number of species has more than doubled since Baker's monograph in the "Flora Capensis" and many more remain to be treated.

Interest in the genus *Lachenalia* now focuses on several other disciplines. Hybridization has been in progress at the Roodeplaat Horticultural Research Institute for many years, and many magnificent hybrids have been produced. The study of the seed patterns and the sculptures of the seed coats by means of the scanning electron microscope have provided new characters for identification. Major work is being carried out at the Jodrell Laboratory at Kew, United Kingdom on the chromosomes, which is proving to be as interesting and as varied as the species.

Interest in the cultivation of the species has increased in many countries, and several valuable papers on horticultural collections have been produced such as "Notes on the Cultivated Liliaceae, Lachenalia" in "Baileya" by John Ingram in 1966, in the USA, and "The Genus Lachenalia" by Trevor Crosby in "The Plantsman", in 1986, in Great Britain. "The Lachenalia Handbook" by Graham D. Duncan is the latest addition, and with its horticultural notes and large number of colour photographs, with the names to them updated, it should prove to be of value and interest to the many enthusiasts, and useful in filling the gap until the major work appears.

W. F. Barker

Important Features

Lachenalia is a genus of small, bulbous geophytes belonging to the family Liliaceae (or Hyacinthaceae *sensu stricto*), and is most closely related to the endemic Cape genus *Polyxena* Kunth.

a. *Bulb*

The bulb in *Lachenalia* is tunicate, having an outer covering of soft or hard, dry membranous tunics which protect it from drying out and physical injury. The fleshy bulb consists of inner and outer bulb-scales. The outer bulb-scales function as storage organs and contain reserve food material. The inner scales are more leaf-like and protect the central growing-shoot which annually produces leaves and flowers. The bulb-scales are attached to a basal plate, from which adventitious roots arise. The size and shape of the bulb varies from species to species, from the minute bulbs of *L. patula* (5–9 mm diam.) to the usually large, fleshy bulbs of *L. bulbifera* (up to 35 mm diam.).

A feature of the bulbs of many species such as *L. giessii*, *L. nordenstamii* and *L. isopetala* from very dry terrain, is the formation of a distinct fibrous neck from old rigid tunics, which provides added protection during unfavourable conditions.

b. *Foliage*

In most species, the mature bulb produces two leaves, while in others a single leaf is produced, and several have numerous grass-like leaves. The genus exhibits remarkable variety in its leaves; they vary from robust and broad as in certain forms of *L. bulbifera* and *L. undulata* to short and cylindrical as in the succulent leaves of *L. patula*. The foliage is usually produced in an upright or spreading position, but in *L. latifolia* and *L. congesta* for example, the leaves lie flat on the ground, while *L. pusilla*, one of the most unusual species, produces its leaves in a rosette at ground level.

The presence of simple or stellate hairs on the leaves of several species is noteworthy. In *L. trichophylla*, hair is found on the entire upper surface and varies in length according to habitat, while in *L. ameliae* the upper surface may be quite smooth, or covered with very short hairs, or it may have varying numbers of hairs restricted to the leaf-margins. In *L. hirta*, hairs occur mainly on the underside of leaves and on the margins, but sporadic hairs may also occur on the upper surface.

The undulating leaf-margin is another characteristic often encountered in *Lachenalia*. *L. undulata* is an obvious example, but it is a very variable species, and the leaf-margins of certain forms are much less undulate than others, and in some instances, leaves may not undulate at all.

Spotting and banding on *Lachenalia* leaves is a conspicuous feature of many species. The colour and density of spots varies with aspect and locality; *L. rubida* for example, growing in full sun, usually has conspicuous purplish spots, while those in shade have spots in shades of green. Similarly, isolated populations of *L. viridiflora* in the Vredenburg district, growing in close proximity, may be heavily spotted or completely unspotted. Spotting in *Lachenalia* usually occurs on upper leaf-surfaces, but sporadic spots may also occur on the undersides.

A wide variety of banding-patterns on the leaf-bases of many species occurs; *L. zebrina* shows very conspicuous, heavy banding, while in *L. macgregoriorum* the bands are narrow and may be interspersed with minute spots.

Numerous species bear pustules on their leaves ranging in size from fairly large, irregularly-scattered ones such as on *L. stayneri* to the small, dense ones on *L. minima*. In *L. pustulata,* pustule size varies with locality, and in certain forms, they may be completely absent.

In the vast majority of species, leaves and flowers are present simultaneously, but in certain species, such as *L. macgregoriorum* and *L. muirii,* the leaves are proteranthous in the wild, but in cultivation, leaves and flowers may occur together.

c. *Inflorescence*

Three different types of inflorescence are encountered in the genus. Firstly the spike, where the flowers are sessile and attached directly to the rachis, as for example in *L. elegans* and *L. muirii* (figure 1a); secondly the subspicate inflorescence, where the flowers are attached to the rachis by very short pedicels, as for example in *L. comptonii* and *L. martinae* (figure 1b) and thirdly the raceme, where the flowers are attached by long pedicels, as for example in *L. aloides* and *L. violacea* (figure 1c). Inflorescence type is not always constant in some species; *L. capensis* for example, may have a spicate or a subspicate inflorescence. A subspicate inflorescence is taken as having a pedicel length of up to 2 mm, and a racemose inflorescence a pedicel length of usually more than 2 mm, but occasionally a maximum length of 2 mm.

d. *Flower, Fruit and Seed*

Flowers range in shape from long and tubular as in *L. bulbifera* to small and campanulate as in *L. bachmannii,* while the position of the flowers on the rachis varies from pendulous to erect.

In all species the perianth is zygomorphic, some obviously so, others only slightly, and consists of three inner and three outer segments. The inner segments usually protrude beyond the outer segments, or they may be almost equal in length and are fused near the base forming a distinct tube, or they are spreading (figure 1d).

In the vast majority of species, a swelling, known as a gibbosity, is present near the apex of the outer perianth segments; they are sometimes large and conspicuous as in *L. margaretae* and *L. montana.*

The six stamens arise from the base of the perianth. They are usually declinate and vary in position from included within to well exserted beyond the tip of the perianth. Included stamens are taken as being shorter than or as long as the perianth, while just protruding stamens are taken as protruding up to 2 mm beyond the tip of the perianth. Shortly exserted stamens are taken as usually protruding slightly more than 2 mm beyond the tip of the perianth, but occasionally protruding to a maximum length of 2 mm. Well-exserted stamens are taken as protruding well beyond 2 mm. In those species where the length of the stamens is not equal, the above refers to the longest stamens. The simple style may be shorter than, as long as or longer than the stamens.

The superior ovary is tri-locular and may contain few to many ovules.

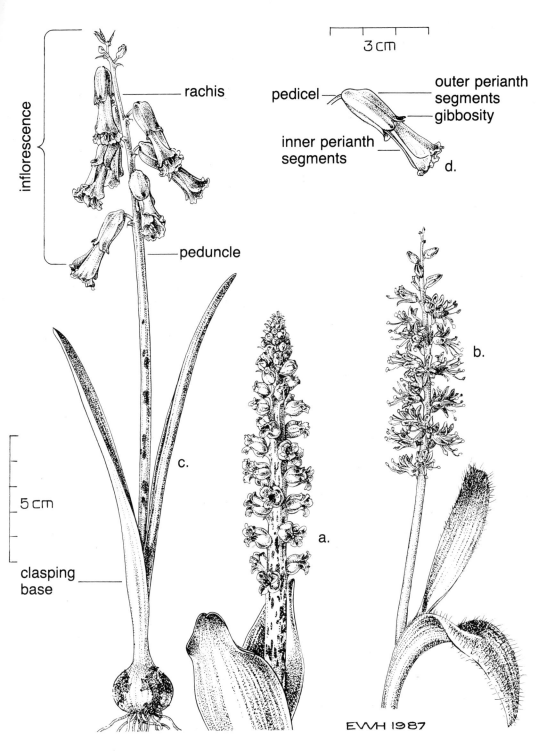

FIGURE 1

General morphology of *Lachenalia*; a. *L. elegans* var. *membranacea* indicating a spicate inflorescence with spreading, urceolate flowers and included stamens; b. *L. comptonii* indicating a subspicate inflorescence with spreading, widely campanulate flowers and well-exserted stamens; c. *L. aloides* var. (Durbanville) indicating a racemose inflorescence with pendulous, tubular flowers and included stamens; d. details of a flower with *L. aloides* var. (Durbanville) as the example.

The fruit is a tri-locular capsule and when ripe, it ruptures length-wise releasing the seeds, which are hard, black and usually shiny and vary considerably in size from the tiny seeds of *L. angelica* (0,7 mm diam.), to the comparatively large ones of *L. isopetala* (2 mm diam.). The morphology of the ripe seed is presently regarded as one of the most important taxonomic characters in *Lachenalia* and has proved useful to Miss W. F. Barker in her revision of the genus. The length of the arillode differs greatly among the species, varying from almost completely absent to comparatively long. The seed coat is smooth and shiny in many species while in some it is wrinkled, or variously patterned.

DISTRIBUTION AND HABITAT

Lachenalia is essentially a genus of the winter rainfall region of southern Africa, with the vast majority of species occurring in the western and south-western Cape, and southern Namaqualand. Several species occur in predominantly summer rainfall regions, while others may be found in areas receiving year round rainfall. It must be mentioned that, in order to gain a more comprehensive idea of its distribution, extensive collecting in many areas is still required, as can be gleaned from the fact that several species are recorded from significantly remote localities, resulting in large "blank" areas in between with no record of material having been collected. Certain presently recognized species have not been seen in the living state for several decades, emphasizing the importance of the few herbarium collections made of these species in the distant past.

The deciduous nature of the genus, inhospitable terrain and the fact that growth and flowering are dependent on sufficient seasonal rainfall are all factors which can be attributed to the as yet incomplete distribution records.

From present records, the most northerly occurring species is *L. giessii* from the Aus district in south-western SWA/Namibia, followed by *L. pearsonii* from the Central Groot Karasberg in the south-east, while several species occur in the southern parts of that country. The range extends into the Richtersveld and throughout Namaqualand, western and south-western Bushmanland, the Knersvlakte, the western and south-western Cape and the Little Karoo. Only a few species are recorded from the Great Karoo, while the southern Cape, particularly in coastal areas, contains numerous species, but the diversity decreases rapidly as one moves eastwards. *L. algoensis* is recorded from as far east as the Transkei, while an as yet undescribed species occurs as far inland as the south-western Orange Free State.

Lachenalia species in the wild occur in a very large range of differing habitats, and consequently species with wide distributions such as *L. bulbifera* and *L. violacea* show much variability in flower colour, size and flowering period.

L. bulbifera for example, with its numerous different forms, begins its flowering season in late April with the short orange blooms of the Still Bay form, followed by the robust bright red west coast forms in May and June. These are followed by various Cape Flats and Peninsula forms in July and August, and finally a September-flowering, orange-red form from the De Hoop Nature Reserve brings its season to an end, the species being in flower for almost six months of the year!

Certain species such as *L. aloides* var. *quadricolor* and *L. viridiflora* are only to be found in the humus-rich depressions and crevices on granite outcrops, while *L. rubida* occurs in nutrient-poor sand, and *L. muirii* favours the limestone flats of the southern Cape coast. The mineral-rich, barren Knersvlakte is host to numerous species, as is the Nieuwoudtville Plateau of considerably higher altitude and rainfall.

Although most species favour open, sunny aspects, a number occur quite naturally in shade, such as *L. margaretae* and *L. orchioides*, while several prefer seasonally inundated flats or marshes, examples being *L. arbuthnotiae* and *L. salteri* respectively.

ENDANGERED SPECIES

Due to the lack of comprehensive distribution knowledge it is difficult to determine the conservation status of many of the species. Several are presently known from just a single locality, such as *L. macgregoriorum* and *L. margaretae*, but they could have a wider distribution, and since they occur in largely inaccessible terrain, they are not in immediate danger. Other species are, however, definitely known to be in an endangered position. Two local endemics on the Cape west coast, *L. mathewsii* and *L. viridiflora* are good examples. *L. mathewsii* was believed to be extinct for several decades until it was recently re-discovered and is today known from a single locality, while *L. viridiflora* is restricted to granite outcrops in the area, and survives mainly on private land.

L. purpureo-caerulea from the Darling district is now endangered due to agricultural activity, but still survives in the Tinie Versfeld Reserve, while *L. polyphylla* from the Boland and *L. arbuthnotiae* from the Cape Flats have been much reduced in numbers, and are in a vulnerable position. Likewise, the numbers of *L. muirii* from the southern Cape are almost certainly dwindling, due to the development in this area.

Several species such as *L. polypodantha* and *L. pearsonii* have not been seen for many years, and their position is quite uncertain.

L. sargeantii, a recent discovery with a very limited known distribution, appears to be stimulated into flowering only after fire, and has not been recorded since 1971.

Although many species are fortunately still common, due either to their prolific nature or wide distribution, it is clear that an increasing number are in need of conservation.

THE SPECIES DESCRIPTIONS

The alphabetical list provided consists of eighty-eight species, with synonyms. Insufficiently known or doubtful species have been omitted due to a lack of living material required for study. The genus is presently estimated to comprise in excess of one hundred and ten species; approximately twenty new species have yet to be formally described and published.

   In this Handbook the Species Descriptions have been arranged alphabetically in two major groups (according to the position of the stamens), and each of these in smaller subgroups (according to the type of inflorescence). *L. hirta* and the very variable *L. contaminata* are exceptions within the above arrangement as they may have included or well-exserted stamens; they have been placed with the former group as is their most usual habit. *L. unifolia* var. *schlechteri* is also an exception in that its pedicels are either very short or absent; it has nevertheless been placed in the subgroup with racemose inflorescences, as is the most usual habit for *L. unifolia*.

KEY TO GROUPS AND SUBGROUPS

The information provided for each species includes the following:
   i. Derivation of specific epithets
   ii. Common names where applicable
   iii. General notes on distribution range
   iv. Details of the foliage, flowers and any other characteristics of particular interest. This information refers to the species in its natural habitat. The colouring of the base of the outer perianth segments is taken here to include the perianth tube; the latter has not been mentioned specifically. Details of seed morphology have been omitted as having little relevance within the scope of this publication.
   v. Horticultural merit of the species
   vi. Flowering period and height range. This information refers to the species in its natural habitat.

. *Lachenalia elegans* var. *flava*
(Ceres)

3b. *Lachenalia elegans* var. *membranacea*
(Clanwilliam)

3c. *Lachenalia elegans* var. *suaveolens*
(in habitat, Nieuwoudtville)

3d. *Lachenalia fistulosa*
(Signal Hill)

4a. *Lachenalia fistulosa*
(Signal Hill)

4b. *Lachenalia framesii*
(Vanrhynsdorp)

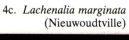

4c. *Lachenalia marginata*
(Nieuwoudtville)

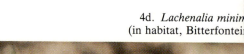

4d. *Lachenalia minim*
(in habitat, Bitterfontei

5a. *Lachenalia muirii*
(Bredasdorp)

5b. *Lachenalia mutabilis*
(Citrusdal)

5c. *Lachenalia mutabilis*
(in habitat, Langebaan)

5d. *Lachenalia mutabilis*
(Springbok)

6a. *Lachenalia namaquensis*
    (Steinkopf)

6b. *Lachenalia orchioides* var. *orchioides*
    (Caledon)

6c. *Lachenalia orchioides* var. *orchioides*
    (Rondebosch)

6d. *Lachenalia orchioides* var. *glauci*
    (Kirstenbosc

vii. Cultivation performance: The allocation here ranges from "poor" to "excellent" and is based on experience gained with the living collection at Kirstenbosch which is located in the winter rainfall region of the Cape; cultivation performance elsewhere may well be at variance.

COMMON NAMES

The popular name "Cape cowslip" has been applied to lachenalias in general for very many years, but it is a rather misleading one as the wild British cowslips (genus *Primula*) bear little resemblance. The locally used South African common names are far more descriptive, such as "viooltjie", which refers to the violin-like "squeaks" produced when *Lachenalia* stems rub together.

THE ILLUSTRATIONS

The illustrations provided should give the reader some idea of the variability in form and flower colour which exists in many species. Illustrations are also provided of some species with particularly interesting or diagnostic leaves, where it was not possible to capture both inflorescence and foliage in a single illustration.

Whenever possible, species were photographed in their natural habitat, while a neutral background was usually used when photographed under cultivation. A general locality is given with each illustration, with "in habitat" added when photographed in the wild.

Out of a total of eighty-eight species included in this publication, eighty-one are illustrated.

HOW TO USE THIS HANDBOOK TO IDENTIFY A SPECIES

Having studied the details concerning stamens and inflorescence under Important Features in the INTRODUCTION,

a. inspect the position of the stamens in a fully open flower (the anthers should be ripe) and determine whether they are:
    i. included, or just protruding beyond the tip of the perianth, or
    ii. shortly exserted to well exserted beyond the tip of the perianth, and turn to the relevant Group under the Species Descriptions, using the KEY on page 12.
b. Now determine which of the three different types of inflorescence is present, and turn to the relevant subgroup. Bear in mind that the type of inflorescence varies in some species.
c. Read through the descriptions for each species in the subgroup, taking particular note of:
    i. number and shape of leaves, as well as markings or other characteristics
    ii. shape and colour combinations of flowers and gibbosities, as well as the position of inner and outer segments
    iii. any other characteristics such as colour of stamens, fragrance etc.

    iv. flowering period and height range

    v. distribution range

The above hints, taken in conjunction with the illustrations, should provide some aid to identification.

It has unfortunately not been possible to illustrate every single species described in this publication; in such instances, extra careful study of the descriptions is recommended for identification.

As already mentioned, the species descriptions refer to plants in their natural habitat. Under cultivation, additional leaves may develop, flower colour may alter, pedicels and stamens may elongate, the flowering period may change and plant height may become exaggerated; the hints on species identification will be more successful when applied to plants in habitat than under cultivation.

It should also be borne in mind that there are approximately twenty undescribed new species not included in this publication, which will account for the reader not being able to make an identification.

CULTIVATION

The cultivation notes set out here are applicable to the areas of Mediterranean-type climate in the Cape Province; it must be borne in mind that lachenalias are not frost-tolerant, requiring protection in susceptible areas. In the Northern Hemisphere, the approximate corresponding month may be obtained by adding or subtracting six months for any particular month mentioned.

All *Lachenalia* species can be successfully cultivated in containers, while only some species are suited to general garden culture.

Container Subjects

Generally speaking, the *Lachenalia* as a container subject is the most rewarding and practical way in which to cultivate it, as pots may easily be moved to the most suitable position, taken indoors during the flowering period or stored away during dormancy.

a. *Aspect*

The aspect for container-grown lachenalias should be sunny and have free air circulation. Pots on a stoep or patio can be arranged together in groups, and flat-dwellers can grow them in window-boxes on a sunny balcony. In areas which receive heavy winter rainfall, the more delicate species are best grown under cover. The specialist grower will be inclined to erect his own structure with benches, open sides and glass-fibre roof, where his ever-expanding collection can be maintained. In Northern Hemisphere countries with extreme winter temperatures, container cultivation in the cool greenhouse is recommended.

b. *Growing medium*

One of the most important factors in container cultivation is the drainage of the growing medium, which must be excellent at all times. Despite the fact that lachenalias in the wild occur in a wide range of soil types with greatly differing pH values, they generally adapt very easily to new soil types under cultivation. The most important component of the growing medium is sand, which should preferably be a medium-grained, washed riversand. For easily cultivated species such as *L. aloides*, *L. pustulata* and *L. unicolor*, a mixture of equal parts riversand and loam or fine compost is ideal. For less easily cultivated species such as *L. ameliae* and *L. violacea*, the amount of loam or compost should be reduced considerably to produce a mixture of one part loam or compost to three parts riversand. At Kirstenbosch, the use of a medium-grained industrial sand in combination with riversand has also been found to give satisfactory results. Difficult species such as *L. ventricosa* and *L. klinghardtiana* should be grown in pure riversand. A layer of stone chips should always be placed over the holes in the bottom of the container to ensure perfect drainage. Pot size and type are important considerations if one intends building up a collection, as uniformity in

15

pot size saves space, and the collection is more easily maintained. Ordinary brown plastic pots are ideal, with 200 mm diamter pots suitable for small to medium-sized species and 250 mm diameter pots for the larger species. Asbestos troughs, hanging-baskets and earthenware pots can also be used, although the latter two tend to dry out very quickly.

c. *Planting*

As soon as temperatures begin to fall after the summer, *Lachenalia* bulbs become active, first with root growth, followed by leaf-shoots. Bulbs should be set out in the autumn, March and April being the most suitable months. The depth of planting will depend on the size of the bulb, but for most species, 2 cm is the correct depth. Species with large bulbs such as *L. bulbifera* can be planted up to 4 cm deep, while the minute bulbs of *L. patula* should be placed about 1 cm deep. The number of bulbs per pot again depends on the size of the bulbs, but lachenalias enjoy being crowded together and give a better display this way. For the larger species, six to eight bulbs to a 250 mm diameter pot, and for smaller species, twelve to fifteen bulbs to a 200 mm diameter pot is recommended.

d. *Watering*

Once planted, pots should be watered well, and then not again until the leaf shoots begin to appear, after which a good soaking once a fortnight is recommended, as opposed to light applications at irregular intervals. Over-watering of container-grown lachenalias will soon lead to rotting, and as a general rule, it is always preferable for the growing medium to be slightly dry than too wet.

Certain species such as *L. anguinea*, *L. latifolia* and *L. stayneri*, despite having perfectly healthy bulbs, occasionally remain dormant throughout a growing season and only become active in the following growing season. Pots of such bulbs are best allowed to dry out and stored.

When in flower, pots can be taken indoors for home decoration, where the delicate fragrance of certain species such as *L. peersii* and *L. fistulosa* can be fully appreciated. While indoors they should be placed in a light, airy position, and watered sparingly. Pots should not remain indoors for more than two weeks, as this results in lanky, weak growth.

Towards the end of spring, as temperatures rise, lachenalias will naturally begin to go dormant, which is indicated by a yellowing of the leaves. Watering should then be withheld completely, and as soon as the leaves have withered, the containers should be placed in a cool dry place and stored.

GARDEN SUBJECTS

The pendulous-flowered *Lachenalia* species and hybrids are popular for bedding and in rockeries where, massed together, they create a brilliant display in winter and spring. There are also a number of species with open-faced flowers suitable for garden culture, and a list of recommended species is provided at the end of this section.

Only certain species are suitable as the vast majority will not survive garden irrigation during their dormant period, and in any event, most species are not showy enough to warrant a place in the garden. The many brightly-coloured varieties of *L. aloides* are popular, as they multiply rapidly, and the flowers are long-lasting. The most readily available *Lachenalia* in the trade is the one sold under the erroneous names of *"L. piersonii"* or *"L. pearsonii"*. It has orange-yellow flowers tipped with maroon, and it is in fact a hybrid raised many years ago in New Zealand. *L. bulbifera* is a robust, striking species suitable for the rockery with very well-drained soil.

The aspect for lachenalias in the garden must be sunny and the soil very well-drained. Drainage in heavy soil can be improved by mixing in large quantities of well-decomposed compost and riversand. Slightly sloping ground is ideal for planting as it allows for maximum water run-off. Groups of lachenalias can be planted in pockets in a rockery, which should be lined with wire-mesh if moles are prevalent. They may also be displayed to great advantage by interplanting with low-growing spring annuals such as *Dorotheanthus bellidiformis* and *Steirodiscus tagetes*. Bulbs should be planted out in groups in March and April at a depth of between two and four cm depending on size, watered well and left alone until leaf-shoots appear, after which thorough soakings every fortnight are recommended if natural precipitation is lacking.

Certain species such as *L. arbuthnotiae*, *L. bulbifera* and *L. aloides* make satisfactory cut-flowers which are long-lasting, and are best arranged in shallow vases containing not more than two cm of water. It is advisable to cut the ends of the stems periodically to lengthen the vase-life of the flowers.

If one is not prepared to lift and replant *Lachenalia* bulbs in the garden each year, then the site chosen for planting should receive as little water as possible during the dormant season to lessen the likelihood of rotting.

LIST OF SPECIES RECOMMENDED FOR GARDEN CULTURE

L. aloides	*L. pallida*
L. arbuthnotiae	*L. peersii*
L. bulbifera	*L. purpureo-caerulea*
L. contaminata	*L. pustulata*
L. elegans	*L. reflexa*
L. fistulosa	*L. rubida*
L. juncifolia	*L. salteri*
L. liliflora	*L. splendida*
L. mathewsii	*L. trichophylla*
L. namaquensis	*L. unicolor*
L. orchioides	*L. unifolia*
L. orthopetala	*L. viridiflora*

PROPAGATION

Seed

Most species produce an abundance of seed each season, and under uncontrolled conditions will hybridize freely. In order to obtain pure seed, pot-isolation and hand-pollination are necessary. Fine hair brushes or cotton-wool-buds can be used to transfer pollen from the anthers of a flower to the stigma of a flower on another plant, provided the material is not a clone.

The hard, black, shiny seeds are best sown in autumn (March to May) in deep seed-trays or pots, in a sterilized medium of equal parts fine compost or loam, and riversand. Seed must be sown thinly to prevent overcrowding and allow sufficient room for bulb development, and should be covered with a thin layer of sand and be kept moist. Containers should be placed in a semi-shaded position for their first season, whereafter a sunny position can be chosen. After germination, seedlings should be kept moist but not wet, and a watering-can with a fine rose should be used.

Generally, *Lachenalia* seed germinates very readily, and fresh seed of the majority of species germinates in about eighteen days, while *L. bachmannii* and *L. patula* are usually up in eleven days, but *L. macgregoriorum* and *L. muirii* can take up to six weeks. By contrast seed of *L. margaretae* often germinates very erratically and may at times only germinate in its second season after sowing.

At the end of their second season, seedling bulbs can be lifted, stored and potted-up in autumn. Many species flower readily in their second season, but as a general rule, first flowering can be expected in the third season. *Lachenalia* seed is easily stored and remains viable for at least five years at room temperature.

Offsets, Bulbils and Stolons

Offsets are side-bulbs which develop out of the mother-bulb, from which they eventually break away, becoming separate bulbs. Not all the species reproduce readily by this method, but for those which do, it is an easy way of increasing one's stock, and one is certain that true-to-type material is being obtained. During the dormant period, offsets which are large enough can simply be removed from the mother-bulb and stored until planting-time in autumn.

Certain species produce small bulblets at, or above ground level, and these are commonly known as bulbils. An example of this is the aptly named *L. bulbifera* of which some forms produce bulbils along the leaf-base margins. These bulbils may simply be removed at the end of the growing season and stored until autumn.

Some species reproduce vigorously by means of stolons, two examples being *L. namaquensis* and *L. moniliformis,* which produce long stolons from the base of the bulb, each carrying a bulbil produced at ground level. *L. aloides* var.

quadricolor, in addition to offsets, often multiplies by producing masses of tiny bulbils on short stolons under the papery bulb-tunics. These bulbils eventually rupture the tunics and become separate bulbs.

LEAF CUTTINGS

Propagation by leaf cuttings is a very satisfactory method of increasing stock of species and of those which don't set seed readily.

Leaves for cutting material are best taken from bulbs which are in active growth, and should be selected from healthy, virus-free plants. In species which produce only one leaf, up to half may be used as cutting-material, whereas in those which produce two or more leaves, an entire leaf may be used. Depending on size, the leaf material can be cut into two or three cross-sections and be placed in a well-drained rooting medium such as equal parts riversand and vermiculite, with the base of the cutting about 1 cm below the surface. The cuttings should be placed in a shaded position and kept slightly moist. After approximately one month, roots and bulblets will begin to form at the base of the cutting. When the original leaf cutting eventually withers, it is time to discontinue watering and inspect and store the bulblets until autumn. It is usually the large-leaved species which produce the most satisfactory results, often producing up to ten sizeable bulblets per cutting, many of which will flower in the following season, for example *L. arbuthnotiae* and *L. purpureo-caerulea.* The more delicate species such as *L. bachmannii* and *L. hirta* may take a further season before flowering.

In many cases, by taking leaf-cuttings as opposed to sowing, one effectively reaches the flowering-stage a year sooner, and the offspring is exactly true to type.

MICRO-PROPAGATION

This is the process of propagating new plants by culturing tiny portions of the *Lachenalia* leaf in an artificial medium, under sterile conditions. Although this is not a method of propagation which would normally be used by the home gardener, it is as well to mention that lachenalias can be successfully propagated by this method, as has been the case at the Horticultural Research Institute at Roodeplaat, outside Pretoria, where some striking hybrids have been raised. Apart from the advantage in being able to propagate large numbers of plants within a short period, micro-propagation is also a means of producing virus-free material, which in *Lachenalia* is an important consideration.

FEEDING

In the author's experience, *Lachenalia* species can be successfully grown without any supplementary feeding. The nutritional requirements of species which naturally occur in humus-rich soil, such as *L. aloides* var. *quadricolor,* are catered for under cultivation in the loam or compost used in the growing medium. This is not to say that feeding is not recommended; on the contrary, most species respond very readily. Fertilizers with a high potash but low nitrogen content can be used. Bone meal is a recommended, slow-acting organic fertilizer which can be mixed into the growing-medium or sprinkled on the surface. A liquid fertilizer with a relatively low nitrogen content can be used at fortnightly intervals.

7a. *Lachenalia orchioides* var.
glaucina (Devil's Peak)

7b. *Lachenalia undulata*
(Bitterfontein)

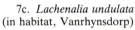

7c. *Lachenalia undulata*
(in habitat, Vanrhynsdorp)

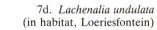

7d. *Lachenalia undulata*
(in habitat, Loeriesfontein)

8a. *Lachenalia bowkeri*
(Port Elizabeth)

8b. *Lachenalia buchubergensis*
(Richtersveld)

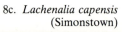

8c. *Lachenalia capensis*
(Simonstown)

8d. *Lachenalia kliprander*
(Klipran

Lachenalia longibracteata
(Saldanha)

9b. *Lachenalia longibracteata*
(in habitat, Piketberg)

9c. *Lachenalia variegata*
(Darling)

9d. *Lachenalia trichophylla*
(in habitat, Klawer)

10a. *Lachenalia trichophylla*
(Klawer)

10b. *Lachenalia trichophylla*
(Garies)

10c. *Lachenalia trichophylla*
(Garies)

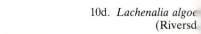
10d. *Lachenalia algoe*
(Riversd

PESTS AND DISEASES

Under cultivation, lachenalias are subject to various pests and diseases, and the following measures are suggested for their control.

PESTS

a. *Mealybug*

These tiny sucking insects can cause serious damage as they multiply and spread rapidly. They are pinkish-white in colour and attack the bases of leaves and may also occur on the bulb's basal plate and under the bulb tunics. Mealybug attack results in deformed leaves, accompanied by yellowish flecking and the eventual disintegration of the bulb. A contact insecticide such as Dursban is recommended for control of leaf-attack, while a systemic insecticide such as Protekta A can be applied as a drench for bulb-attack. To control mealybug on loose bulbs during storage, dusting with Extermathion powder is suggested.

b. *Aphids*

These cause stunting damage, and are also carriers of viral disease, and must be controlled immediately infestation is noticed. Malathion or Lebaycid can be used for control.

c. *Caterpillars, slugs and snails*

Caterpillars can be troublesome on leaves and often cause damage to ripening seed-capsules, but are easily controlled with Karbaspray. Slugs and snails may cause extensive damage, particularly to the leaves of the larger species, for which a recognized snail bait can be used.

DISEASES

a. *Damping-off*

The damping-off fungus causes weakening and collapse in *Lachenalia* seedlings, and it is prevalent when seed has been sown too thickly, resulting in poor aeration especially after watering. It can be prevented by sowing seed thinly in sterilized soil and spraying with Kaptan as soon as infection is noticed. Drenching the soil with a Benlate solution is a recommended alternative.

b. *Rust*

An unsightly rust (*Uromyces lachenaliae*) may occur on *Lachenalia* leaves which takes the form of small pustules which rupture and disperse masses of orange spores, eventually destroying an entire leaf. Thorough weekly applications of Dithane M45 is suggested for control.

c. *Rotting*

Fungal rotting of *Lachenalia* bulbs is more often than not the result of overwatering and poor soil aeration. A well-drained, sterilized growing medium and correct watering procedure will reduce the likelihood of rotting.

d. *Virus*

Lachenalias under cultivation are very susceptible to virus diseases, and the symptoms are easily recognized as yellowish-brown streaking on leaves, and deformed flowers or leaves. It does not always lead to immediate death of the plant, but such material is best destoyed before healthy plants become infected, as there is no reliable cure at present.

DISEASE AND WEED CONTROL BY SOIL STERILIZATION

Soil sterilization is an effective and recommended method of controlling soil-borne fungi and weed-seeds.

Heat treatment

For the home gardener growing just a few pots of *Lachenalia*, ready-mixed, damp soil can be sterilized in the oven in shallow trays at a temperature of 93° C for one hour. Alternatively, containers filled with ready-mixed soil can be sterilized with a thorough application of boiling water.

Chemical treatment

For larger quantities of soil, or beds in the garden, sterilization with Basamid is recommended, which involves applying the chemical to damp soil, digging-in thoroughly, and covering with plastic for about two weeks.

The above-mentioned chemicals are poisonous and potentially dangerous; they are to be applied with great care.

STORAGE METHODS DURING DORMANCY

Lachenalias in the wild state undergo a dormant period during the hot, dry summer months, and survive these conditions in the form of a bulb. Under cultivation, this period of dormancy must be simulated, and as soon as the leaves have withered, bulbs in the garden may be lifted, and containers removed to a cool, dry place for storage. This is an ideal time to inspect bulbs for insect or fungal damage, treat accordingly and store until autumn. Bulbs can be stored loosely in paper packets, or they can be placed in dry sand, peat or wood-shavings.

Species in pots which do not need to be divided and which one is sure are healthy, can simply be stored as they are.

Finally, it is important that different species are properly labelled during storage.

SOURCES OF SUPPLY

All *Lachenalia* species are protected by law, and may not be collected in the wild without the necessary permits.

For those wishing to start their own collection of *Lachenalia* species, the following sources of supply are recommended.

By joining the Botanical Society of South Africa, one can take advantage of their annual catalogue of surplus seed supplied by the National Botanic Gardens, which usually has a selection of *Lachenalia* species on offer. At the Society's annual Plant Sale one can purchase bulbs of a number of lachenalias.

Membership of the Indigenous Bulb Growers Association of South Africa (IBSA), entitles one to obtain the addresses of specialist bulb nurseries, from which a wide selection of species can be acquired.

The Wild Flower Gardens at Caledon and Clanwilliam sometimes have *Lachenalia* seed on offer.

1. The Administrative Manager
   Botanical Society of South Africa
   Kirstenbosch
   CLAREMONT
   7735
   RSA

2. The Secretary
   IBSA
   PO Box 141
   WOODSTOCK
   7915
   RSA

3. Caledon Wild Flower Garden
   PO Box 24
   CALEDON
   7230
   RSA

4. Clanwilliam Wild Flower Garden
   PO Box 24
   CLANWILLIAM
   8135
   RSA

THE SPECIES DESCRIPTIONS

GROUP 1   **Stamens included or just protruding beyond the tip of the perianth**
(stamens shorter than, as long as, or protruding up to 2 mm
beyond the tip of, the perianth)

Subgroup 1a.   **Inflorescence spicate** (pedicels absent)

Lachenalia ameliae *W.F. Barker*   (Plate 1a–b)

ameliae: after Mrs A. Amelia Mauve, author of many botanical publications

A dwarf species from the Ceres and Montagu districts.

One or two short, lanceolate to ovate-lanceolate, unmarked leaves are
produced which may either be smooth or hairy; the upper leaf surface may be
covered in very short hairs, or hairs of varying length, sometimes confined to the
leaf margins. The lower leaf surface is tinged with maroon and the clasping base
may be spotted or banded with magenta. The inflorescence consists of sessile,
urceolate-oblong flowers; the outer perianth segments may be pale yellow or
greenish-yellow with pale green gibbosities, and sometimes with a narrow
maroon or purple central stripe. The protruding inner segments are pale yellow
or greenish-yellow, the tips spreading or recurved and may sometimes be tinged
with magenta or distinctly marked with purple.

L. ameliae is one of the most desirable dwarf species, ideally suited to pot
culture.

Flowering period	August–September
Height	40–115 mm
Cultivation performance	Good

Lachenalia arbuthnotiae *W. F. Barker*   (Plate 1c)

arbuthnotiae: after Miss I. Arbuthnot, botanical assistant at the Bolus and
Compton herbaria until 1945

Probably once a very common species, this Cape Flats endemic is now
restricted to isolated remnants of the fynbos in this area, occurring in seasonally
inundated, low-lying ground.

The bulb produces one or two lanceolate, leathery leaves which may be plain
green or maroon, or green with dense spotting, mainly on the upper surface. The
inflorescence is a dense spike of bright yellow, oblong-shaped flowers; the outer
perianth segments have pale green gibbosities and the inner segments protrude.
Each flower is subtended by a white, narrow bract, conspicuously so from the
middle of the inflorescence upwards. Flowers are sweetly scented and fade to
dull red as they mature.

Horticulturally, *L. arbuthnotiae* must be regarded as one of the most
desirable species, suited to both pot and garden culture. It is also a long-lasting
cut flower.

Flowering period	August–October
Height	180–400 mm
Cultivation performance	Excellent

Lachenalia carnosa *Bak.* (Plate 1d–2b)

carnosa: fleshy leaves

A widely distributed, well-known species from Namaqualand which used to be known as *L. ovatifolia* L. Guthrie.

It is often a robust plant, normally producing two bright green, ovate to broadly-ovate, lanceolate leaves which have depressed longitudinal veins on the upper surface. The foliage is usually unmarked, but can have green or brown pustules above. The inflorescence consists of numerous sessile, urceolate-oblong flowers; the outer perianth segments are dull white and have green or maroon gibbosities, while the protruding inner segments are white with broad mauve or magenta tips.

This showy species requires excellent drainage, and is well suited to pot culture.

Flowering period   August–September
Height   80–250 mm
Cultivation performance   Good

Lachenalia concordiana *Schltr.* ex *W.F. Barker* (Plate 2c)

concordiana: after the town Concordia, Namaqualand

A seldom seen species which has been collected very infrequently; once from Concordia and subsequently from near Springbok, Garies and Calvinia.

The bulb produces a single linear-lanceolate, conduplicate leaf which is banded with green in the upper half of the clasping base and with dull purple in the lower portion. The inflorescence consists of widely-campanulate, sessile flowers, usually arranged in whorls of three. The outer perianth segments are pale blue at the base, shading to very pale cream and green, and have dark green gibbosities and recurved tips. The protruding inner segments are dull white with green keels and are also recurved.

An interesting species for the specialist grower.

Flowering period   September
Height   60–200 mm
Cultivation performance   Good

Lachenalia congesta *W.F. Barker* (Plate 2d)

congesta: flowers congested together

A very distinctive, usually dwarf species from the Sutherland and Calvinia districts.

The frequently deep-seated bulb produces two ovate to oblong dark green, prostrate leaves (under cultivation leaves tend to grow in a raised position). The leaf margin is maroon, and scattered dark markings may or may not be present on the upper surface, while the lower leaf surface is tinged with maroon. The very dense inflorescence consists of sessile, strongly scented, oblong-campanulate flowers; the outer perianth segments are very pale blue at the base, shading to white or very pale yellow, and have green gibbosities and recurved tips. The protruding inner segments have green markings near the tips and are recurved.

The peduncle is usually not visible above ground-level in the wild.

L. congesta would only appeal to the specialist grower and is best suited to pot culture.

Flowering period	June–August
Height	80–140 mm
Cultivation performance	Excellent

Lachenalia elegans *W.F. Barker*

elegans: elegant inflorescences

A very variable, usually montane species favouring sandy, often moist conditions. It is most frequent in the Nieuwoudtville district and its range extends south to the Cedarberg and western Karoo. Four named varieties are listed here.

i)   var. **elegans**

This, the typical variety, is only known to occur near Nieuwoudtville.

One or two bright green, lanceolate leaves are produced which may be plain or spotted, and have thickened brown margins. The inflorescence consists of numerous oblong-urceolate, suberect flowers; the outer perianth segments are bright blue at the base, shading to rose and have brown gibbosities, while the protruding inner segments are white with a pale pink spot near the tips. The other varieties of this species are all easily distinguished from var. *elegans* in having spreading, urceolate flowers.

This variety is not yet in common cultivation, but has definite horticultural potential.

Flowering period	October
Height	180–240 mm
Cultivation performance	Excellent

ii)   var. **flava** *W.F. Barker*   (Plate 3a)

flava: yellow flowers

A most attractive variety from the Elands Kloof, Ceres and Karoo Poort districts.

Usually a single glaucous, lanceolate to ovate-lanceolate leaf is produced which has dark green blotches on the upper surface, and a maroon, crisped margin. The inflorescence consists of urceolate, spreading flowers; the outer perianth segments are bright yellow with maroon tips and pale green gibbosities, while the protruding inner segments are bright yellow with a dark maroon zone near the tips and a narrow white, membranous margin.

This is the earliest flowering of all the varieties, ideally suited to pot culture with excellent drainage.

Flowering period	July–August
Height	150–250 mm
Cultivation performance	Excellent

ii) var. **membranacea** *W.F. Barker* (Plate 3b)

membranacea: refers to the broad membranous margin of the inner perianth segments

Occurs commonly from Nieuwoudtville to Clanwilliam.

The bulb produces one or two lanceolate to ovate-lanceolate, glaucous or bright green leaves which are usually spotted above with darker green or maroon. The inflorescence consists of urceolate, spreading flowers; the outer perianth segments are pale or dark yellow with green gibbosities and the protruding inner segments are pale green with a brownish zone near the tips, and a broad white, membranous margin.

This floriferous, attractive variety is recommended for both pot and garden culture.

Flowering period	August–September
Height	150–200 mm
Cultivation performance	Excellent

iv) var. **suaveolens** *W.F. Barker* (Plate 3c)

suaveolens: refers to its distinctive sweet fragrance, reminiscent of carnations

Occurs in the Nieuwoudtville and Clanwilliam districts.

It produces one or two lanceolate to ovate-lanceolate, green leaves which may or may not be spotted with dark green above. The inflorescence consists of urceolate, scented, spreading flowers; the outer perianth segments are pale blue or green at the base, shading to pink, and to dark maroon in the upper third, and have dark maroon gibbosities. The protruding inner segments are dark maroon in the upper half and have narrow white, membranous margins.

An attractive variety recommended for both pot and garden culture.

Flowering period	August–September
Height	100–270 mm
Cultivation performance	Excellent

Lachenalia fistulosa *Bak.* (Plate 3d–4a)

fistulosa: refers to the hollow formed by the surrounding perianth segments

A heavily fragrant species associated with rocky mountain slopes occurring in the Piketberg, Tulbagh and Worcester districts, the Cape Peninsula and as far east as Caledon. It was previously known as *L. convallariodora* Stapf.

The bulb produces two lorate leaves which may be plain or heavily spotted above. The usually slender peduncle bears a spike of oblong-campanulate flowers; the outer perianth segments vary in shades of cream, yellow, blue, lilac or violet, and have pale brown gibbosities. The protruding inner segments are clearly recurved in the upper half and are generally pale-coloured.

L. fistulosa should be grown mainly for its heavy, sweet scent and is highly recommended for both pot and garden culture.

Flowering period	September–October
Height	80–300 mm
Cultivation performance	Excellent

Lachenalia framesii *W.F. Barker*   (Plate 4b)

framesii: after Mr P. Ross-Frames, collector and cultivator of succulent plants

A dainty dwarf species with an attractive contrasting flower colour, occurring mainly in the Vanrhynsdorp district on the Knersvlakte, but as far north as Komaggas and often growing in large colonies, favouring flat, sandy areas.

One or two linear, plain green leaves are produced which may be erect or recurved, and usually have distinctly undulate margins. The flowers are sessile, urceolate-oblong and borne on a very slender, light green peduncle. The outer perianth segments are greenish-yellow with yellowish-brown or green gibbosities and the protruding inner segments are bright purple or magenta at their tips and recurved.

L. framesii is floriferous and is one of the most desirable species as pot subjects.

Flowering period	July–August
Height	90–150 mm
Cultivation performance	Good

Lachenalia marginata *W.F. Barker*   (Plate 4c)

marginata: leaves have thickened margins

A fairly common species in the Nieuwoudtville, Vanrhynsdorp and Clanwilliam districts, usually occurring singly in heavy or sandy ground.

The plant is easily recognized by its single glaucous, usually ovate leaf which has an undulate thickened margin. The upper leaf surface has pale brown markings and the clasping base has distinctive maroon bands. The inflorescence consists of sessile, oblong-cylindrical flowers; the outer perianth segments are very pale blue at the base, shading to white or pale yellow, and have dark brown gibbosities, while the protruding inner segments are greenish-yellow. The peduncle is often conspicuously swollen just below the inflorescence.

L. marginata would appeal mainly to the specialist grower.

Flowering period	July–August
Height	110–300 mm
Cultivation performance	Good

Lachenalia minima *W.F. Barker*   (Plate 4d)

minima: very small plants

An early flowering, usually dwarf species only known from the Bitterfontein district in southern Namaqualand, frequenting moist clay flats.

11a. *Lachenalia bachmannii*
(Piketberg)

11b. *Lachenalia contaminata*
(in habitat, Darling)

11c. *Lachenalia contaminata*
(Riviersonderend)

11d. *Lachenalia martinae*
(Clanwilliam)

12a. *Lachenalia maximiliani*
(Wuppertal)

12b. *Lachenalia reflexa*
(in habitat, Claremont)

12c. *Lachenalia schelpei*
(Calvinia)

12d. *Lachenalia dehooper*
(in habitat, Bredasdo

a. *Lachenalia isopetala*
(Calvinia)

13b. *Lachenalia liliflora*
(Bellville)

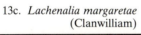

13c. *Lachenalia margaretae*
(Clanwilliam)

13d. *Lachenalia mediana* var. *mediana*
(Observatory)

14a. *Lachenalia mediana* var. *rogersii*
      (Porterville)

14b. *Lachenalia orthopetala*
      (Durbanville)

14c. *Lachenalia orthopetala*
      (Durbanville)

14d. *Lachenalia palli*
      (in habitat, Durbanvil

It produces two lanceolate or ovate-lanceolate leaves which have depressed longitudinal veins and are usually densely covered with small pustules on the upper surface. The inflorescence is a dense spike of pale yellow, oblong-campanulate flowers; the outer perianth segments have greenish gibbosities and recurved tips, and the protruding inner segments have green central keels and are also recurved. The flowers are borne in an upright position and fade to dull red as they mature.

An interesting species for the specialist grower.

Flowering period	June
Height	20–170 mm
Cultivation performance	Good

Lachenalia muirii *W.F. Barker* (Plate 5a)

muirii: after Dr J. Muir, collector of plants in the Riversdale district

A late-flowering, mainly coastal species favouring limestone hills and flats from Bredasdorp to Still Bay, and inland to Riversdale.

It belongs to the small group with proteranthous leaves; in the wild, the plant comes into flower after the one or two linear leaves have withered. The inflorescence consists of very distinctive sessile, urceolate-oblong flowers; the outer perianth segments are pale blue at the base shading to white or very pale pink and have large dark brown or maroon gibbosities, while the inner segments each have a maroon or brown central keel and protrude conspicuously.

L. muirii is highly recommended for pot culture.

Flowering period	October–December
Height	100–250 mm
Cultivation performance	Excellent

Lachenalia mutabilis *Sweet* (Plate 5b–d)

mutabilis: refers to the changing colour of the inflorescence

Common in the Clanwilliam district, this extremely variable species occurs throughout Namaqualand, south as far as Piketberg, Langebaan on the west coast, inland as far as Worcester and south to Riviersonderend.

The relatively small bulb normally produces one lanceolate, often erect leaf, which may be plain green or glaucous, and which may be faintly spotted or occasionally banded with maroon on the clasping base. The leaf margins are often crisped and the peduncle is frequently swollen just below the inflorescence, the latter being a dense spike of oblong-urceolate flowers. In the typical forms, such as from the Clanwilliam district, the outer perianth segments are pale blue at the base, shading to dull white, and have dark brown gibbosities, while the protruding inner segments are dark yellow with a brown marking near the tips. The apex of the rachis is a bright electric blue and consists of numerous sterile, pedicelled flowers. In the less typical forms, such as from Namaqualand, the flowers are not as colourful, with greenish inner segments and the peduncle is more swollen.

Good forms of this species are very desirable, but require excellent drainage under cultivation.

Flowering period   July–September
Height   100–450 mm
Cultivation performance Good

Lachenalia namaquensis Schltr. ex *W.F. Barker*   (Plate 6a)
namaquensis: from Namaqualand

This attractive, aptly named species occurs in the Steinkopf and Springbok districts and is also recorded from the Richtersveld, frequenting exposed rocky habitats, often forming large colonies.

The bulb is characterized by reproducing rapidly by means of long stolons produced at the base of the bulb which terminate in bulbils at ground level. One or two linear, lanceolate leaves are produced which are unmarked and distinctly conduplicate. The inflorescence consists of urceolate-oblong, sessile flowers; the outer perianth segments are very pale blue at the base, shading to magenta and have greenish-purple or maroon gibbosities, while the two upper protruding inner segments are white with magenta tips, and the lower magenta.

L. namaquensis is very floriferous and highly recommended for both pot and garden culture.

Flowering period   August–October
Height   80–230 mm
Cultivation performance Excellent

Lachenalia orchioides *(L.) Ait.*
orchioides: apparently refers to its orchid-like scent

A very variable, usually montane species, often growing in shade and comprising two varieties:

i)   var. **orchioides**   (Plate 6b–c)
common names: groenviooltjie, wild hyacinth

This is the most widespread variety, occurring commonly on flats or mountain slopes from the Clanwilliam district to the Cape Peninsula, inland as far as Worcester, and eastwards to Swellendam. It was previously known as *L. glaucina* Jacq. var. *pallida* Lindl.

The one or two leathery, lanceolate or lorate leaves are not pustulate, but may or may not be spotted with brown on the upper surface. The inflorescence consists of sweetly-scented, oblong-cylindrical flowers; the outer perianth segments are pale blue at the base, shading to greenish-yellow or creamy-yellow and have green gibbosities, while the much longer inner segments are often conspicuously recurved at their tips. Flowers fade to dull red as they mature.

There are some particularly fine forms of this variety recommended for both pot and garden culture.

Flowering period	August–October
Height	80–400 mm
Cultivation performance	Excellent

ii)   var. **glaucina** *(Jacq.) W.F. Barker* . (Plate 6d–7a)

glaucina: blue-grey flowers

common name: blouviooltjie

Previously known as *L. glaucina* Jacq., this variety differs from var. *orchioides* in its flower colouring in shades of blue and it is restricted to the eastern slopes of Table Mountain. It is also not as strongly scented as the typical variety.

The outer perianth segments are blue at the base, shading to purple with dark purple gibbosities, or they may be entirely blue with dark blue gibbosities. There exists a small-flowered form of var. *glaucina* which occurs on the slopes below Devil's Peak; its outer perianth segments are blue at the base, shading to mauve, pink or violet, and have dark brown gibbosities. Its inner segments are recurved at the tips, whitish at the base and shade to dark mauve, pink or violet.

L. orchioides var. *glaucina* is one of the most desirable members of the genus and deserves wide horticultural attention.

Flowering period	August–October
Height	100–400 mm
Cultivation performance	Excellent

Lachenalia undulata *Masson* ex *Bak.*   (Plate 7b–d)

undulata: undulating leaf margins

A very variable species with a wide distribution, occurring mainly in Namaqualand and the Knersvlakte, its range extending from the Richtersveld to as far south as the Klawer district.

The bulb produces two unbanded, lanceolate to ovate-lanceolate leaves which usually have undulate margins. Leaf colour varies from bright green to glaucous, and in the latter forms, sporadic dark brown spots may be present on the upper surfaces. The leaves are also characterized in having depressed longitudinal veins on the upper surface. The inflorescence of sessile, oblong-campanulate flowers is borne on a plain green peduncle. The outer perianth segments are dull white, tinged with green and pale blue, each with a dark brown gibbosity. The protruding inner segments are white, each with a dark brown central marking.

Horticulturally, *L. undulata* has little to commend it and would only appeal to the specialist grower.

Flowering period	May–June
Height	100–300 mm
Cultivation performance	Good

Lachenalia verticillata *W.F. Barker*

verticillata: flowers arranged in whorls

A very rarely collected species at present recorded from the Springbok district and south-western Bushmanland.

It produces a single lanceolate, conduplicate leaf which is glaucous, marked with purple on the lower surface and has an undulate margin. The clasping leaf base is banded with purple. The inflorescence consists of sessile, urceolate-shaped flowers arranged in whorls of three. The outer perianth segments are pale blue and the protruding inner segments are white with magenta-coloured recurved tips.

The species is as yet unknown in cultivation, but may have potential as a pot subject when material becomes available.

Flowering period	September
Height	100–250 mm
Cultivation performance	Unknown

Subgroup 1b: **Inflorescence spicate or subspicate** (pedicels absent or up to 2 mm long)

Lachenalia bowkeri *Bak.*   (Plate 8a)

bowkeri: after Col. J. H. Bowker, collector of plants and butterflies from Transkei, Natal and Zululand

A poorly known eastern Cape species recorded from the Port Elizabeth and Riebeek-East districts.

It produces one or two glaucous, lanceolate leaves which are conduplicate and may be plain or minutely spotted with magenta on the clasping base. The inflorescence consists of oblong-campanulate flowers; the outer perianth segments are speckled with blue near the base and have dull purple tips and green gibbosities. The protruding inner segments are white with brown or purple speckled keels.

Not a particularly attractive species, being more of interest to the specialist grower.

Flowering period	August
Height	100–260 mm
Cultivation performance	Good

Lachenalia buchubergensis *Dinter*   (Plate 8b)

buchubergensis: after the Buchuberg, SWA/Namibia

A rare, dwarf species from the Richtersveld and south-western corner of SWA/Namibia.

The relatively large bulb produces a single lanceolate, falcate leaf which is spotted on the lower surface, and banded with maroon on the clasping leaf base. The few-flowered inflorescence consists of tubular, sessile or almost sessile flowers; the outer perianth segments are olive-green or blue-green, each with a greenish-brown gibbosity, while the slightly protruding inner segments are bright green with purple or maroon tips.

L. *buchubergensis* would be of interest to only the most avid collector, but is as yet unknown in cultivation.

Flowering period	July
Height	60–70 mm
Cultivation performance	Unknown

Lachenalia capensis *W.F. Barker* (Plate 8c)

capensis: from the Cape Peninsula

A fragrant, late-flowering species restricted to the Cape Peninsula, usually occurring on mountain slopes.

One or two lanceolate or lorate, pale-green leaves are produced and may be plain or have brown spots of varying sizes on the upper surface. The peduncle too, can be plain or have brown markings. The inflorescence consists of white or cream, oblong-cylindrical flowers; the outer perianth segments are very pale blue at the base, sometimes shading to very pale yellow at the tips and have pale yellow gibbosities. The inner segments protrude, and flowers fade to brownish-pink as they mature.

L. *capensis* should be grown mainly for its very sweet fragrance.

Flowering period	September–October
Height	150–250 mm
Cultivation performance	Excellent

Lachenalia kliprandensis *W.F. Barker* (Plate 8d)

kliprandensis: after the town Kliprand

A most beautiful, very distinctive species from south-western Bushmanland. It has been collected on just two occasions and plants occur singly in deep red sand.

It produces two ovate, prostrate leaves which have depressed longitudinal veins and numerous dark brown or green pustules on the upper surface. The sturdy peduncle is either very short or not visible above ground-level and bears a many-flowered inflorescence of urceolate-oblong flowers. The outer perianth segments are white with brownish-green gibbosities and pale magenta tips, and the protruding inner segments are white with broad magenta tips.

A very desirable pot subject.

Flowering period	August–September
Height	100–200 mm
Cultivation performance	Excellent

Lachenalia longibracteata *Phillips* (Plate 9a–b)

longibracteata: flowers subtended by long bracts

A variable, rather inconspicuous *Lachenalia* from the Piketberg, Vredenburg and Malmesbury districts, favouring rocky outcrops, stony mountain slopes and moist flats.

It has one or two lanceolate, often leathery leaves which may be plain or spotted above. The inflorescence is usually a spike, but may also be subspicate, and consists of oblong-shaped flowers; the outer perianth segments can be pale blue at the base, shading to yellow, or they can be entirely blue with scattered darker spots, and each has a brown or green gibbosity. The protruding inner segments are pale yellow or cream-coloured. The flowers are characterized in each being subtended by a conspicuous, usually long, narrow bract.

The species is insufficiently attractive to be recommended for general cultivation.

Flowering period   July–September
Height   70–350 mm
Cultivation performance   Excellent

Lachenalia variegata *W.F. Barker* (Plate 9c)

variegata: variegated speckles occur on outer perianth segments

Due to its drab colouring, this species is rather inconspicuous in the wild. It occurs from the Clanwilliam district as far south as the Cape Peninsula, but is most frequently seen along the west coast in deep sand.

On sandy flats the species is often robust, while on mountain slopes it is more stunted and has paler flowers with stamens becoming exserted. The bulb normally produces a single, erect or spreading leaf which varies from lanceolate to lorate, and has thickened, often undulate margins. The leaf is glaucous on the upper surface and finely spotted with blue-purple below and on the clasping base and peduncle. Occasionally the clasping base may be banded with maroon. The inflorescence is usually a spike, but may also be subspicate. It bears numerous oblong-campanulate flowers; the outer perianth segments have variegated speckles of green, blue or purple, with a brown or green gibbosity at the apex. The protruding inner segments are green with white margins.

The species is rather short-lived under cultivation and requires excellent drainage.

Flowering period   August–October
Height   100–400 mm
Cultivation performance   Fair

Subgroup 1c: **Inflorescence spicate, subspicate or racemose** (pedicels absent, up to 2 mm long or longer)

Lachenalia trichophylla *Bak.* (Plate 9d–10c)

trichophylla: leaves covered with hairs

It would appear that the distribution of this very distinctive species is divided into two zones; a southern zone extending from Citrusdal to Vanrhynsdorp, and inland to Nieuwoudtville and Wuppertal, and a northern zone from Nuwerus to Kamieskroon.

The bulb produces a single heart-shaped leaf (a second leaf may develop under cultivation) and is covered with stellate hairs of varying length on the upper surface and margin, depending on locality; the forms occurring in the northern zone, previously known as *L. massonii* Bak., are characterized in having both short and long stellate hairs, while those from the southern zone have short, hardly visible stellate hairs. The forms occurring in the southern zone have sessile flowers with several abortive, pedicelled flowers at the apex, while the plants from the northern zone have lax inflorescences with flowers borne on short or long pedicels. The inflorescence consists of oblong-cylindrical flowers which vary in shades of pale yellow, and may be flushed with pink on the outer perianth segments; the latter have green gibbosities and the inner segments protrude.

   L. trichophylla makes a free-flowering, attractive pot and garden subject and is highly recommended.

Flowering period	August–September
Height	80–200 mm
Cultivation performance	Excellent

Subgroup 1d: **Inflorescence subspicate** (pedicels up to 2 mm long)

Lachenalia algoensis *Schonl.*   (Plate 10d)

algoensis: after Algoa Bay, Port Elizabeth

   This well-known, very variable *Lachenalia* is widely distributed; inland in the Worcester district, and along the coast from Bredasdorp as far east as Transkei.

   The one or two leaves vary from linear to lanceolate or lorate; light green markings may or may not be present on the upper surface, and the peduncle can be plain or spotted. The inflorescence is of the subspicate type; the characteristically erect, cylindrical-ventricose flowers are carried on very short pedicels. The outer perianth segments are pale green, often faintly tinged with blue at the base, and each has a green or brown gibbosity. The inner segments are pale yellow or green and protrude conspicuously. Flowers fade to dull red as they mature, and certain forms are pleasantly scented.

   L. algoensis would appeal mainly to the specialist grower as its flowers are rather short-lived.

Flowering period	July–September
Height	60–300 mm
Cultivation performance	Excellent

Lachenalia bachmannii *Bak.*   (Plate 11a)

bachmannii: after Dr F. Bachmann, naturalist and explorer

   Occurs in the Piketberg, Malmesbury and Stellenbosch districts, favouring seasonally inundated, marshy ground.

It is sometimes confused with *L. contaminata*, but the latter species has numerous grass-like leaves whereas *L. bachmannii* has only two leaves which are linear, conduplicate and unmarked. The inflorescence consists of campanulate, white flowers; the outer perianth segments each have a brown or dark red gibbosity, and the slightly longer inner segments each have a dark red or brown marking near the tip.

Not a particularly attractive species, being more of interest to the specialist grower.

Flowering period	August–September
Height	150–300 mm
Cultivation performance	Excellent

Lachenalia contaminata *Ait.*   (Plate 11b–c)

contaminata: thought to refer to the maroon-coloured floral markings
common name: wild hyacinth

A very variable, widespread species in wet, sandy or heavy ground ranging from Citrusdal, south throughout the south-western Cape and as far as Bredasdorp.

It belongs to the small group of species with numerous grass-like leaves; the number, colour and length differs from locality to locality. They are usually unmarked, semi-terete, channelled above and can be erect or lie horizontally. The slender peduncle is usually marked with dark maroon. The dense inflorescence consists of campanulate to widely-campanulate white flowers; the outer perianth segments each have a dark maroon or brown gibbosity, while the slightly longer inner segments have a dark maroon central stripe near the tips. In most forms the stamens are included, but in several they are clearly exserted.

The species does extremely well under cultivation and needs to be massed together for effect.

Flowering period	August–October
Height	60–250 mm
Cultivation performance	Excellent

Lachenalia martinae *W.F. Barker*   (Plate 11d)

martinae: after Miss B. E. Martin, a former Kirstenbosch horticulturist

A localized species in the Clanwilliam district occurring in rocky habitats in heavy soil.

The distinctive, single leaf is recognized by its undulate margin, ovate-lanceolate shape and the conspicuous maroon bands on the clasping base. The inflorescence consists of oblong-campanulate flowers; the outer perianth segments are dull white, minutely spotted with pale blue or grey and have pale greenish-brown gibbosities, while the protruding inner segments have a reddish-brown marking near the tips.

Lachenalia rubida
(in habitat, Saldanha)

15b. *Lachenalia rubida*
(in habitat, Cape Point)

15c. *Lachenalia viridiflora*
(Vredenburg)

15d. *Lachenalia youngii*
(in habitat, Knysna)

16a. *Lachenalia zeyheri*
(Ceres)

16b. *Lachenalia aloides* var. *aloides*
(Malmesbury)

16c. *Lachenalia aloides* var. *aurea*
(Bainskloof)

16d. *Lachenalia aloides* var. *quadric*
(Darl

Lachenalia aloides var. *quadricolor*
(in habitat, Langebaan)

17b. *Lachenalia aloides* var. *vanzyliae*
(Porterville)

17c. *Lachenalia aloides* var.
(Red Hill)

17d. *Lachenalia aloides* var.
(Chapman's Peak)

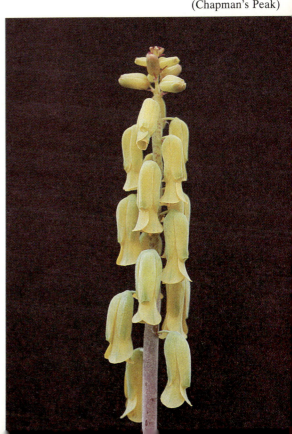

18a. *Lachenalia aloides* var.
(Cape Point)

18b. *Lachenalia aloides* var.
(Durbanville)

18c. *Lachenalia aloides* var.
(Riebeek-Kasteel)

18d. *Lachenalia aloides*
(Piketb

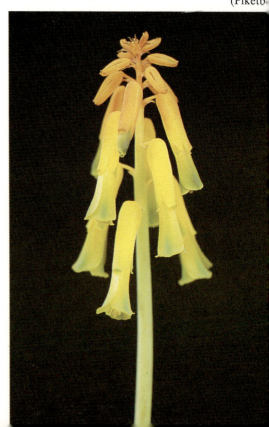

19a. *Lachenalia angelica*
(Hondeklipbaai)

19b. *Lachenalia bolusii*
(Jenkinskop)

19c. *Lachenalia bolusii*
(Jenkinskop)

19d. *Lachenalia bulbifera*
(Saldanha)

20a. *Lachenalia bulbifera* (in habitat, Vredenburg)

20b. *Lachenalia bulbifera*
(in habitat, Bloubergstrand)

20c. *Lachenalia bulbi*
(Bredasd

. *Lachenalia bulbifera*
(Still Bay)

21b. *Lachenalia giessii*
(Rosh Pinah)

21c. *Lachenalia hirta* var. *hirta*
(Citrusdal)

21d. *Lachenalia hirta* var. *hirta*
(Citrusdal)

22a. *Lachenalia hirta* var. *hirta*
(Malmesbury)

22b. *Lachenalia leomontana*
(Swellendam)

22c. *Lachenalia namibiensis*
(Rosh Pinah)

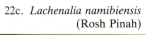

22d. *Lachenalia pat*
(Moedverlo

23a. *Lachenalia peersii*
(Hermanus)

23b. *Lachenalia rosea*
(Gansbaai)

23c. *Lachenalia sargeantii*
(Bredasdorp)

23d. *Lachenalia unifolia* var. *unifolia*
(Gouda)

24a. *Lachenalia unifolia* var. *unifolia*
(Gouda)

24b. *Lachenalia unifolia* var. *unifolia*
(Darling)

24c. *Lachenalia unifolia* var. *wrightii*
(Darling)

24d. *Lachenalia splend*
(Vanrhynsdo

Lachenalia splendida
(Bitterfontein)

25b. *Lachenalia ventricosa*
(Klawer)

25c. *Lachenalia barkeriana* (Bitterfontein)

26a. *Lachenalia comptonii*
(Matjiesfontein)

26b. *Lachenalia comptonii*
(Matjiesfontein)

26c. *Lachenalia physocaulos*
(Robertson)

26d. *Lachenalia gil*
(Piketb

Horticulturally the species has little to commend it and would only be of interest to the specialist grower.

Flowering period	July–August
Height	100–250 mm
Cultivation performance	Fair

Lachenalia maximiliani *Schltr.* ex *W.F. Barker* (Plate 12a)

maximiliani: after Maximilian Schlechter, trader and plant collector of the Port Nolloth district

A dainty species from the Wuppertal district frequenting steep mountain slopes.

The bulb is characterized by being surrounded by several layers of brown tunics which enclose numerous bulbils. Usually a single lanceolate, conduplicate leaf is produced which is yellowish-green in colour and unmarked. The inflorescence of oblong-campanulate flowers is borne on a slender peduncle; the flowers at the base are sessile, while those in the upper portion have very short pedicels. The outer perianth segments are pale blue at the base, shading to white and have brown or reddish-brown gibbosities. The protruding inner segments are white with bluish-green keels, and are recurved at their tips and fade to dark magenta as they mature.

The species is not particularly attractive and would probably only appeal to the specialist grower.

Flowering period	July–August
Height	100–200 mm
Cultivation performance	Good

Lachenalia reflexa *Thunb.* (Plate 12b)

reflexa: leaves bent downwards

A common, usually dwarf species frequenting seasonally inundated flats from Malmesbury to Franschhoek and the Cape Peninsula.

The one or two bright green or glaucous leaves vary from lanceolate to lorate, are usually reflexed, and can be plain or heavily spotted on the upper surface. The leaf margins are often thickened and usually undulate. The peduncle is usually very short and almost completely clasped by the leaf bases. The greenish-yellow, cylindrical-ventricose flowers are borne in an upright position; the outer perianth segments have green or greenish-yellow gibbosities and the inner segments protrude. Flowers fade to dull red as they mature.

Good forms of this species are well worth growing and it makes a pleasing, early flowering pot or garden subject.

Flowering period	June–August
Height	30–190 mm
Cultivation performance	Good

Lachenalia schelpei *W.F. Barker*   (Plate 12c)

schelpei: after Prof E. A. Schelpe, author of numerous botanical works on
          orchids and ferns

This species is at present only known to occur on the Hantamsberg in the
Calvinia district, and has been collected very infrequently.

Its two lanceolate leaves have large green markings on both the upper and
lower surfaces, and the loosely clasping base is banded with maroon. The
inflorescence bears numerous scented, oblong-urceolate flowers; the outer
perianth segments are white with greenish-brown gibbosities and the protruding
inner segments are white with a green marking near the tips. Each flower is
subtended by a long bract which is particularly conspicuous during the bud stage.

The species is insufficiently attractive to be recommended for general
cultivation, and would only appeal to the specialist grower.

Flowering period   June–July
Height   100–225 mm
Cultivation performance   Good

Subgroup 1e: **Inflorescence subspicate or racemose** (pedicels up to 2 mm long or
          longer)

Lachenalia dasybotrya *Diels*

dasybotrya: shaggy inflorescences

A poorly known species from the Calvinia and Nieuwoudtville districts
occurring on red clay flats.

The bulb is characterized by being covered with layers of fibrous tunics,
produced into a neck, while the two lanceolate glaucous leaves are unmarked
and can vary from narrowly to broadly ovate. The inflorescence consists of
oblong-campanulate flowers; the outer perianth segments vary from whitish to
greenish-yellow, with or without pale blue bases, and have brown or green
gibbosities. The protruding inner segments are white with brown or green
markings.

L. dasybotrya is as yet unknown in cultivation, but will probably have
potential as a pot subject when material becomes available.

Flowering period   August–October
Height   50–100 mm
Cultivation performance   Unknown

Lachenalia dehoopensis *W.F. Barker*   (Plate 12d)

dehoopensis: after De Hoop Nature Reserve

A localized species in the Bredasdorp district found in flat, sandy areas.

It produces two linear leaves which are channelled above and banded with
green on the lower surface and with maroon and magenta on the clasping base.
The inflorescence consists of oblong-campanulate flowers; the outer perianth

segments are pale blue at the base, shading to cream, and have reddish gibbosities, while the protruding inner segments are cream with broad red keels.

The species is as yet unknown in cultivation and would probably only appeal to the specialist grower.

Flowering period	August–September
Height	80–160 mm
Cultivation performance	Unknown

Lachenalia isopetala *Jacq.*   (Plate 13a)

isopetala: petals of equal length

This very distinctive *Lachenalia* is found in the Nieuwoudtville, Calvinia, Middelpos and Sutherland districts, favouring red stony ground.

The bulb is characterized by the many cartilaginous tunics surrounding it, which extend into a straw-like neck. In the wild state, the two arching, unmarked glaucous leaves are proteranthous, having withered by the time the flowers open. The inflorescence consists of oblong-shaped flowers of which the outer perianth segments are a dull yellowish-cream with a maroon or brown central band which broadens at the apex. The equal, or very slightly longer inner segments are cream-coloured with a broad maroon or brown zone near the apex.

L. isopetala is one of the late flowering species and makes an interesting pot subject for the specialist grower.

Flowering period	October–November
Height	100–300 mm
Cultivation performance	Excellent

Lachenalia liliflora *Jacq.*   (Plate 13b)

liliflora: funnel-shaped flowers, as in certain *Lilium* species

This late-flowering, attractive species has a limited distribution in the Paarl, Bellville and Durbanville districts, favouring hilly slopes.

Two lanceolate, plain green leaves are produced which are usually densely pustulate on the upper surface. The inflorescence consists of oblong-campanulate flowers; the outer perianth segments are white with brownish gibbosities and the slightly longer inner segments are white with dark magenta tips.

L. liliflora multiplies well under cultivation and makes a very satisfactory pot and garden subject.

Flowering period	September–October
Height	100–200 mm
Cultivation performance	Excellent

Lachenalia margaretae *W.F. Barker*   (Plate 13c)

margaretae: after Mrs Margaret Thomas, plant propagator at Kirstenbosch
          Botanic Garden

This late-flowering, very distinctive dwarf species is at present only known to occur on the Pakhuis Pass near Clanwilliam, favouring shady rock ledges.

The bulb produces a single (or occasionally two under cultivation) lorate leaf which may be plain or heavily spotted above. The short inflorescence consists of campanulate flowers; the outer perianth segments are white with conspicuous brown gibbosities and the very slightly longer inner segments have brown markings near the tips.

L. margaretae is an interesting species for the specialist grower.

Flowering period	October–December
Height	30–120 mm
Cultivation performance	Excellent

Lachenalia mediana *Jacq.*

mediana: intermediate, apparently referring to its intermediate resemblance between *L. pallida* and *L. orchioides*

A variable species with a fairly wide distribution extending from Porterville to the Cape Peninsula and east as far as Caledon. It comprises two varieties.

i)   var. **mediana**   (Plate 13d)

In most forms, two leaves are produced which are lanceolate and plain green; in certain forms the clasping leaf-base may have a maroon-coloured zone. The inflorescence consists of oblong or oblong-campanulate, pale opalescent flowers. The outer perianth segments are pale blue at the base shading to dull white or grey, and each has a green or purplish gibbosity. The protruding inner segments are dull white with a green or purple marking near the tips.

ii)   var. **rogersii** *(Bak.) W.F. Barker*   (Plate 14a)

rogersii: after Rev. W. M. Rogers, collector of plants, mainly in the southern Cape

This variety occurs from Durbanville to Porterville. It differs from var. *mediana* in usually having a single, broader leaf blade with undulate to crisped margins, and the clasping base is irregularly banded with dark maroon to magenta markings. The colouring of its flowers is also more varied, ranging in shades of blue to pink. It used to be known as *L. unifolia* Jacq. var. *rogersii* Bak.

L. mediana is a floriferous, highly recommended pot subject.

Flowering period	August–September
Height	200–400 mm
Cultivation performance	Good

Lachenalia orthopetala *Jacq.*   (Plate 14b–c)

orthopetala: straight petals

Occurs in the Piketberg, Malmesbury and Durbanville districts, but due to farming activity is becoming rather scarce. It favours moist flats and often grows in large colonies.

It belongs to the group with many grass-like leaves; they are deeply channelled above and may be plain or have green or brown spots on the upper surface. The slender peduncle is dark maroon and bears a dense inflorescence; the flowers are white, oblong-campanulate and face upwards. The outer perianth segments each have a dark maroon gibbosity, and the slightly longer inner segments have a dark maroon marking at their tips. Certain forms develop a central, pale maroon stripe on both the inner and outer perianth segments.

L. orthopetala is highly recommended for both pot and garden culture.

Flowering period   September–October
Height   90–270 mm
Cultivation performance Excellent

Lachenalia pallida *Ait.*   (Plate 14d)

pallida: pale-coloured flowers

One of the most common species, *L. pallida* is still to be seen flowering in its thousands in spring, favouring moist, heavy ground in the Piketberg, Malmesbury, Wellington and Stellenbosch districts. It also occurs in the north-western Cape Peninsula and in isolated patches along the west coast to Darling.

One or two unmarked, lanceolate leaves are produced which may or may not be covered with pustules on the upper surface. The inflorescence consists of numerous oblong-campanulate flowers; flower colour ranges from cream to various shades of yellow. The outer perianth segments each have a brown or green gibbosity at the apex, and the inner segments protrude. Flowers fade to dull red as they mature.

Only certain colour forms of this species are sufficiently attractive to be recommended for general cultivation.

Flowering period   August–October
Height   120–300 mm
Cultivation performance Excellent

Lachenalia rubida *Jacq.*   (Plate 15a–b)

rubida: ruby-red flowers

common names: sandkalossie; bergnaeltjie

One of the most attractive and well-known members of the genus. It is the earliest flowering of all the species, frequenting flats, mountain slopes and dunes, often growing within reach of the sea spray. Its distribution is mainly coastal, and extends from north of Hondeklipbaai on the west coast to the Cape Peninsula and eastwards as far as George. It occurs inland as far as Vanrhynsdorp.

The one or two lanceolate or lorate leaves may be plain or spotted with dark purple or various shades of green, mainly on the upper surface. Sporadic spots may also occur on the lower surface. A noteworthy characteristic of the species, due to its usually early flowering nature, is the appearance of flower buds soon after the first autumn rains, before the leaves have fully developed. The inflorescence consists of pendulous, cylindrical flowers. The outer perianth segments vary in colour from bright pink to ruby-red, or intensely spotted with ruby-red on a pale yellow background, and have yellowish-green or pinkish-red gibbosities. The much longer inner segments may be spotted or have purple tips with white markings.

L. rubida deserves wide horticultural attention and is ideally suited to pot culture as well as the rock garden.

Flowering period   March–July
Height   60–250 mm
Cultivation performance   Excellent

Lachenalia viridiflora *W.F. Barker*   (Plate 15c)

viridiflora: green-flowered

This strikingly beautiful species has a limited distribution in the Vredenburg district where it is restricted to humus-rich shallow depressions on granite outcrops. It is usually a dwarf species in nature and blooms early in the season.

The two lanceolate, pale green leaves have longitudinal impressed veins above and may be plain or darkly spotted. Leaves are occasionally pustulate. The inflorescence consists of cylindrical-ventricose flowers; the outer perianth segments vary in colour from viridian green to blue-green or turquoise, and have a viridian green central stripe and gibbosities, while the protruding inner segments have whitish tips and a viridian green central stripe.

Horticulturally, *L. viridiflora* should be regarded as one of the most desirable species, being early flowering, very floriferous and attractive.

Flowering period   May–July
Height   80–200 mm
Cultivation performance   Excellent

Lachenalia youngii *Bak.*   (Plate 15d)

youngii: after Mrs E. M. Young, mycologist and plant collector in the southern
          and eastern Cape

A southern Cape species occurring in coastal areas from Mossel Bay to Humansdorp.

The bulb produces two narrow, lanceolate, unmarked leaves which are channelled above, and loosely clasp the base of the peduncle. The inflorescence consists of campanulate-shaped flowers; the outer perianth segments are pale blue at the base, shading to pink and have purplish-pink gibbosities, while the very slightly protruding inner segments are white with a central dark pink stripe.

The species merits horticultural attention but is unfortunately not yet in cultivation.

Flowering period   July–November
Height   70–300 mm
Cultivation performance   Unknown

Lachenalia zeyheri *Bak.* (Plate 16a)

zeyheri: after Carl L. Zeyher, eighteenth century plant collector

A delicate species, usually associated with very wet, marshy ground, occurring in large colonies in the Ceres district.

The bulb produces one or two linear, semi-terete, plain green leaves which are channelled above. The inflorescence is a dense, narrow raceme of widely campanulate white flowers; the outer perianth segments have reddish-brown or green gibbosities, with similar coloured markings on the tips of the inner segments. The inner and outer segments are almost equal in length. The flowers fade to dull reddish-pink as they mature.

L. zeyheri is well suited to pot culture and bulbs should be massed together for effect.

Flowering period   September–October
Height   60–200 mm
Cultivation performance Excellent

Subgroup 1f: **Inflorescence racemose** (pedicels usually longer than 2 mm)

Lachenalia aloides *(L.f.) Engl.*

aloides: flowers resemble those of an aloe

The most popular and colourful member of the genus. Previously known as *L. tricolor* Jacq.f., it is a widely-distributed, complex species consisting of many different forms. It is almost always associated with rocky habitat and occurs in coastal areas from Lamberts Bay to the Cape Peninsula, east to Bredasdorp, and inland as far as the Worcester district.

Four validly named varieties are listed here as well as four as yet unnamed varieties. A further listed variety, previously known as *L. tricolor* Jacq.f. var. *luteola* Jacq. has still to be validly published under *L. aloides*.

L. aloides is easily recognized by its lanceolate or lorate, usually heavily marked leaves, and its inflorescence of pendulous, cylindrical flowers in which the inner perianth segments protrude conspicuously.

The species has been used extensively in hybridization.

i)   var. **aloides** (Plate 16b)
Occurs in the Malmesbury district on exposed granite slopes.

The two lanceolate, glaucous leaves may be plain, but are usually heavily marked with green or purple on the upper surface. The outer perianth segments are yellow, or reddish-orange at the base shading to yellow, and have bright green gibbosities. The inner segments are yellow shading to pale green, and have wide red tips.

This variety is very similar to var. *quadricolor*, but the latter differs in having its inner segments tipped with purplish-maroon and the base of its outer perianth segments is always reddish-orange. These two varieties grow in association in the Malmesbury district, and intermediate forms do occur.

L. aloides var. *aloides* is not yet in common cultivation but will, no doubt, be popular when material becomes available.

Flowering period	July
Height	150–260 mm
Cultivation performance	Excellent

ii)   var. **aurea** *(Lindl.) Engl.*   (Plate 16c)

aurea: golden-yellow flowers

common names: geelklipkalossie; geelviooltjie

This striking, well-known variety is known only from the Bainskloof mountains between Worcester and Wellington.

The two lanceolate or lorate leaves may be plain or have purplish blotches or spots on the upper surface. The inflorescence of golden-yellow flowers is borne on a dark maroon peduncle; the outer perianth segments have slightly darker golden-yellow or pale green gibbosities, and the inner segments protrude conspicuously.

Horticulturally, this is one of the most desirable varieties within the species. Bulbs multiply rapidly, and the highly attractive flowers are long lasting.

Flowering period	September–October
Height	60–250 mm
Cultivation performance	Excellent

iii)   var. **quadricolor** *(Jacq.) Engl.*   (Plate 16d–17a)

quadricolor: each flower comprises four different colours

common names: vierkleurtjie; vierkleurkalossie

Found in colonies in humus-rich crevices of granite outcrops in the Malmesbury, Darling and Langebaan districts. It also occurs very sporadically in the Cape Peninsula on north-facing slopes of Signal Hill.

The two lanceolate, glaucous leaves may be plain, but are usually covered with green or maroon blotches on the upper surface. This variety is characterized in reproducing rapidly by means of bulbils, produced on short stolons from the base of the bulb. The inflorescence consists of very distinctive, four-coloured flowers. The outer perianth segments are reddish-orange at the base and most of their length, shading to yellow and have bright green gibbosities. The inner segments are yellow or yellowish-green and have wide purplish-maroon tips. A similar variety (see vi) occurs in the Cape of Good Hope Nature Reserve.

This is the most colourful variety and is ideally suited to pot culture as well as the rock garden.

Flowering period	July–August
Height	90–200 mm
Cultivation performance	Excellent

iv)   var. **vanzyliae** *W.F. Barker*   (Plate 17b)

vanzyliae: after Mrs L. van Zyl, who first introduced it to Kirstenbosch in 1927

This most unusually coloured variety occurs in the Piketberg, Porterville, Elands Kloof, Cedarberg and Twenty Four Rivers mountain ranges.

It produces one or two lanceolate, ovate-lanceolate or lorate leaves which may be unmarked, but which are usually densely marked with purplish-brown on the upper surface. Conspicuous white bracts subtend the pendulous, cylindrical flowers; the outer perianth segments are pale blue at the base, shading to white, and have green or yellowish-green gibbosities, while the protruding inner segments are yellowish-green with white margins.

This highly desirable variety often becomes robust under cultivation and multiplies rapidly.

Flowering period   September–October
Height   50–260 mm
Cultivation performance Excellent

v) var. (Cape Peninsula)   (Plate 17c–d)

This variety was previously known as *L. tricolor* Jacq.f. var. *luteola* Jacq. and has still to be validly published under *L. aloides*. It occurs in the southern Cape Peninsula.

It produces two lanceolate, glaucous leaves which are usually densely marked with purplish-brown on the upper surface. The outer perianth segments are usually pale yellow shading to green and have green or greenish-yellow gibbosities, while the protruding inner segments are greenish-yellow. The tip of the inflorescence often consists of several bright reddish-orange, sterile flowers, subtended by conspicuous red bracts.

A very floriferous, often late-flowering variety.

Flowering period   August–October
Height   150–280 mm
Cultivation performance Excellent

vi) var. (Cape Point)   (Plate 18a)

Known only from the Cape of Good Hope Nature Reserve, occurring in exposed rocky areas near the sea.

It produces two lanceolate, pale green leaves which have purplish-brown spots on the upper surface. This variety is very similar to var. *quadricolor*, but differs in having somewhat narrower flowers, and the bases of the outer perianth segments in the lower part of the inflorescence usually lack the reddish-orange colouring of var. *quadricolor*. It also flowers much later.

This variety is not as easily cultivated as the others and requires excellent drainage.

Flowering period   October
Height   150–250 mm
Cultivation performance Fair

vii) var. **(Durbanville)**   (Plate 18b)

Occurs on exposed rocky outcrops in the Durbanville district.

It produces two narrow-lanceolate, pale green leaves which are usually spotted with darker green on the upper surface. The outer perianth segments are bright yellow and have bright green gibbosities, while the much longer inner segments are bright yellow, with or without reddish-maroon tips.

This variety is the earliest flowering within the species and makes a very satisfactory pot and garden subject.

Flowering period	May–June
Height	150–250 mm
Cultivation performance	Excellent

viii) var. **(Riebeek-Kasteel)**   (Plate 18c)

Occurs on rocky slopes near Riebeek-Kasteel.

The one or two leathery, lanceolate or lorate leaves are pale green or glaucous, and are usually marked with green or purplish-brown on the upper surface. Both the outer and inner perianth segments are greenish-yellow and the gibbosities are green. The tip of the inflorescence consists of several red or orange, sterile flowers.

A relatively short-growing variety, ideally suited to pot culture.

Flowering period	July–September
Height	100–150 mm
Cultivation.performance	Excellent

ix) var. **(Piketberg)**   (Plate 18d)

A most attractive variety from the Piketberg mountains.

It produces two lanceolate, bright green leaves which may or may not be marked with purplish-brown on the upper surface. The inflorescence consists of particularly long, pendulous flowers; the outer perianth segments are tinged with orange at the base, shading to yellow and have yellowish-green gibbosities, while the much longer inner segments are yellow with green tips and keels. The tip of the inflorescence consists of several bright reddish-orange, sterile flowers.

Horticulturally this variety deserves wide attention and is suited to both pot and garden culture.

Flowering period	September
Height	150–310 mm
Cultivation performance	Excellent

Lachenalia angelica *W.F. Barker*   (Plate 19a)

angelica: the whole plant in flower resembles a little angel

This dainty dwarf species is at present only known from its type locality near Hondeklipbaai on the Cape west coast.

27a. *Lachenalia haarlemensis*
(Avontuur)

27b. *Lachenalia klinghardtiana*
(Alexander Bay)

27c. *Lachenalia latifolia*
(Heidelberg)

27d. *Lachenalia latifolia*
(Heidelberg)

28a. *Lachenalia mathewsii*
(Vredenburg)

28b. *Lachenalia purpureo-caerulea*
(Darling)

28c. *Lachenalia pusilla*
(Clanwilliam)

28d. *Lachenalia pusi*
(Auro

29a. *Lachenalia salteri*
(Elim)

29b. *Lachenalia anguinea*
(Komaggas)

29c. *Lachenalia duncanii*
(in habitat, Kliprand)

29d. *Lachenalia esterhuysenae*
(in habitat, Clanwilliam)

30a. *Lachenalia juncifolia* var. *juncifolia*
(Still Bay)

30b. *Lachenalia juncifolia* var. *campanulata*
(Riversdale)

30c. *Lachenalia macgregoriorum*
(Nieuwoudtville)

30d. *Lachenalia macgregorioru*
(Nieuwoudtvill

It produces a single ovate to ovate-lanceolate leaf which may be prostrate or raised and is covered with minute, stellate hairs on the upper surface. Under cultivation, leaves tend to grow in an erect position towards the end of the growing season. The lower leaf surface is suffused with dark maroon. The inflorescence consists of several white, widely-campanulate flowers; the outer perianth segments have very pale green gibbosities and the slightly protruding inner segments have pale yellow keels.

L. angelica is one of the most attractive dwarf species and is ideally suited to pot culture. It requires excellent drainage.

Flowering period   October
Height   60–95 mm
Cultivation performance   Good

Lachenalia bolusii *W.F. Barker*   (Plate 19b–c)

bolusii: after Dr Harry Bolus, plant collector and author of "Orchids of South Africa"

A delicate species occurring in rocky habitats from the Richtersveld to as far south as Clanwilliam.

The bulb produces a single dark green or glaucous, lanceolate to ovate-lanceolate or lorate leaf which is conspicuously banded with maroon on the underside of the tightly clasping base. The inflorescence is a lax raceme of several narrowly campanulate pendulous flowers; the outer perianth segments are pale blue at the base, shading to white with brown gibbosities, and the slightly longer inner segments are white with a reddish-brown marking near the apex.

A very attractive species, well suited to container cultivation with excellent drainage.

Flowering period   August–September
Height   100–350 mm
Cultivation performance   Good

Lachenalia bulbifera *(Cyrillo) Engl.*   (Plate 19d–21a)

bulbifera: bulbils form on leaf-base margins

common name: rooinaeltjie

This is the most robust and striking member of the genus, and used to be well known by its old name *L. pendula* Ait. It is a very widespread and variable species, occurring mainly on dunes and rocky outcrops from west of Klawer and along the coast as far east as Mossel Bay.

The usually large, fleshy bulb produces one or two narrowly to broadly ovate, lanceolate or lorate leaves which can be unmarked, but which are often heavily spotted on the upper surface and peduncle. The inflorescence consists of cylindrical, pendulous flowers which vary in length and colour from locality to locality; the form occurring at Still Bay on the south coast has relatively short, orange flowers, whereas those on the west coast and Cape Flats have long, orange or red flowers in various shades. The outer perianth segments have dark red or brown gibbosities and the slightly longer inner segments usually have green tips, flanked by two purple zones.

This is horticulturally one of the most desirable species, suited to both pot and garden culture.

Flowering period   April–September
Height   80–300 mm
Cultivation performance   Excellent

Lachenalia giessii *W.F. Barker*   (Plate 21b)

giessii: after Mr J. W. Giess, prolific collector of plants in SWA/Namibia

A dwarf species, at present only known to occur in the south-western corner of SWA/Namibia, where it is fairly plentiful in red sand.

The bulb is characterized in being covered with layers of rigid brown tunics, produced into a neck. Two unmarked, glaucous leaves are produced which can vary considerably in shape and length; they may be linear, lanceolate or lorate. The inflorescence consists of narrowly campanulate to widely campanulate white flowers; the outer perianth segments each have a green or reddish-purple gibbosity, while the slightly longer inner segments each have a green or reddish-purple marking near the tips. Stamens tend to elongate under cultivation.

L. giessii has definite horticultural potential as a pot subject, being both floriferous and attractive. It requires excellent drainage.

Flowering period   August–September
Height   60–160 mm
Cultivation performance   Good

Lachenalia hirta *(Thunb.) Thunb.*

hirta: hairy leaves

A common species in the Clanwilliam, Citrusdal and Piketberg districts, its distribution extending from Namaqualand as far south as Malmesbury. It comprises two varieties.

i)   var. **hirta**   (Plate 21c–22a)

It is characterized by its single linear leaf which is covered in stiff hairs on the margin and lower surface. The clasping leaf-bases are conspicuously banded with maroon. The inflorescence is a delicate raceme of oblong-campanulate flowers, borne on very long pedicels. In the typical colour forms, the outer perianth segments are light blue at the base and shade to pale yellow, with dark brown gibbosities. The protruding inner segments are pale yellow with a dark brown marking at the tips. The form occurring at Malmesbury is quite robust, and its perianth segments are dull blue-grey.

ii)   var. **exserta** W.F. Barker

exserta: refers to the well-exserted stamens

This variety occurs in the southern end of the distribution range from Clanwilliam to Moorreesburg. It differs from var. *hirta* in its shorter, more campanulate flowers, and the stamens are well exserted.

L. hirta is an outstanding pot subject and highly recommended.

Flowering period   August–September
Height   100–300 mm
Cultivation performance   Excellent

Lachenalia leomontana *W.F. Barker*   (Plate 22b)

leomontana: after the Leeurivierberg

This delicate, late-flowering species is at present only known from the Leeurivierberg in the Swellendam district.

The bulb is characterized in forming numerous white bulbils at its base, and it produces a single lorate or lanceolate leaf which may be plain green or densely spotted with dark purple on the upper surface. The inflorescence consists of oblong-campanulate, pure white flowers; the outer perianth segments have pale green gibbosities and the protruding inner segments have recurved tips.

L. leomontana is one of the loveliest species and is ideally suited to pot culture.

Flowering period   October–November
Height   100–300 mm
Cultivation performance   Excellent

Lachenalia namibiensis *W.F. Barker*   (Plate 22c)

namibiensis: after Namibia/SWA

A very attractive dwarf species from south-western Namibia/SWA.

The bulb is characterized in being covered with layers of brown or black cartilaginous scales, produced into a short neck. One or two falcate, conduplicate leaves are produced; they are unmarked, glaucous and have a depressed midrib and minutely ciliate margins. The inflorescence consists of numerous widely campanulate, white flowers; the outer perianth segments have pinkish-green gibbosities and the protruding inner segments have dark pink keels.

This very floriferous species becomes quite robust under cultivation and is highly recommended for pot culture.

Flowering period   August
Height   50–100 mm
Cultivation performance   Excellent

Lachenalia patula *Jacq.*   (Plate 22d)

patula: outspread flowers

A very distinctive, dwarf species occurring in large colonies on quartz hillsides in the Klawer, Vredendal and Nuwerus districts. It was previously known as *L. succulenta* Masson ex Bak.

It is characterized by its two short, fleshy semi-terete leaves which are channelled above and reddish-maroon in colour. The bulb is very small and covered with black cartilaginous scales. The inflorescence consists of large, widely campanulate flowers, borne on long pedicels, and flower colour varies from white to pale pink. The outer perianth segments each have a brownish-pink gibbosity, while the much longer inner segments have a dark pink central stripe.

L. patula requires excellent drainage under cultivation and is an attractive, floriferous pot subject.

Flowering period   September–October
Height   60–150 mm
Cultivation performance   Good

Lachenalia pearsonii *(Glover) W.F. Barker*

pearsonii: after Prof H. H. W. Pearson, first Director of the National Botanic
          Gardens

Originally described as *Scilla pearsonii* Glover, this little-known *Lachenalia* has been collected on just one occasion when the type gathering was made during the Percy Sladen Memorial Expedition to the Great Karasberg, SWA/ Namibia, in 1913.

Two linear leaves are produced, and the bulb is surrounded with numerous rigid tunics, forming a distinct neck. The inflorescence consists of small, widely campanulate white flowers which have brownish-blue markings at their tips.

The name *L. pearsonii* has for many years been incorrectly used for the plant which is readily available in the trade; the latter is in fact a New Zealand raised hybrid with pendulous orange-yellow flowers and maroon tips. The true *L. pearsonii* is as yet unknown in cultivation.

Flowering period   January
Height   50–100 mm
Cultivation performance   Unknown

Lachenalia peersii *Marloth* ex *W.F. Barker*   (Plate 23a)

peersii: after Mr V. S. Peers, collector of Cape plants, mainly bulbs and
          succulents

A late flowering, very attractive species with a heavy fragrance reminiscent of carnations. It occurs commonly in the Betty's Bay, Hermanus and Caledon districts in full sun or semi-shade, and may often be seen flowering in profusion after fire.

The one or two lorate leaves are unmarked, and vary in colour from maroon in full sun to green in semi-shade. The inflorescence consists of numerous urceolate, white flowers; the outer perianth segments have green or greenish-brown gibbosities, while the protruding inner segments are recurved at their tips. Flowers fade to dull pink as they mature.

Horticulturally a very desirable, floriferous species with a distinctive fragrance.

Flowering period	October–November
Height	150–300 mm
Cultivation performance	Excellent

Lachenalia rosea *Andrews* (Plate 23b)

rosea: rose-pink flowers

A late-flowering, mainly coastal species occurring from the Cape Peninsula as far east as Knysna, and inland as far as Ladismith and Montagu.

Usually just one leaf is produced which is lanceolate and leathery, and may be plain or heavily marked with maroon or brown on the upper surface. The inflorescence consists of oblong-campanulate flowers; the outer perianth segments can vary considerably in colour from shades of blue to rose-pink, or they can be blue at the base shading to rose-pink. They each have a brown or deep pink gibbosity, and the protruding inner segments are rose-pink.

An attractive species which is well suited to container cultivation.

Flowering period	August–December
Height	80–300 mm
Cultivation performance	Good

Lachenalia sargeantii *W.F. Barker* (Plate 23c)

sargeantii: after Mr Percy Sargeant, keen naturalist and photographer of Cape plants

Undoubtedly one of the most unusual and striking members of the genus, *L. sargeantii* is at present only known to occur on the Bredasdorp mountains. It would seem that it requires fire to be stimulated into successful flowering in the wild, and has not been seen in recent years.

The bulb produces two unmarked, linear-lanceolate, conduplicate leaves. The inflorescence consists of pendulous, cylindrical flowers which are borne on long magenta pedicels at the very top of the rachis. The outer perianth segments are cream or pale green and have green or greenish-brown gibbosities, while the much longer inner segments are cream or pale green with a green or greenish-brown marking near the tips.

The species is unfortunately as yet unknown in cultivation, but will no doubt make a desirable pot subject when material becomes available.

Flowering period	November
Height	200–300 mm
Cultivation performance	Unknown

Lachenalia unifolia *Jacq.*

unifolia: single-leaved

A very variable and widespread species, its distribution extending from

Namaqualand to the Cape Peninsula, eastwards to Bredasdorp and inland as far as the Worcester district. Robust plants are often found along the west coast in deep sand. Three varieties are presently recognized within the species:

i)   var. **unifolia**   (Plate 23d–24b)

The relatively small bulb produces a very distinctive, single linear leaf which widens conspicuously in the lower portion, forming a loosely clasping base which is usually distinctly banded with maroon and magenta. The inflorescence consists of oblong-campanulate flowers; the outer perianth segments vary in shades of blue at the base, shading to white, pale yellow or pink, and have brown, greenish-brown or dark pink gibbosities, while the protruding inner segments vary in shades of white.

ii)   var. **schlechteri** *(Bak.) W.F. Barker*

schlechteri: after F. R. Schlechter, botanist and traveller

Occurs from the Kamieskroon district south to Gouda. It differs from the other two varieties in having suberect flowers which are either sessile or borne on very short pedicles. It was previously known as *L. schlechteri* Bak.

iii) var. **wrightii** *Bak.*   (Plate 24c)

wrightii: after Mr C. Wright, nineteenth century botanical collector from America

This variety occurs within the distribution range of var. *unifolia,* but is less common. It differs in its shorter, smaller flowers, often borne on comparatively shorter pedicels.

L. unifolia is best suited to pot culture, but can also be grown in the rock garden with excellent drainage.

Flowering period   August–October
Height   100–350 mm
Cultivation performance   Good

GROUP 2:   **Stamens shortly exserted to well exserted beyond the tip of the perianth** (stamens usually protruding more than 2 mm beyond the tip of the perianth)

Subgroup 2a: **Inflorescence spicate** (pedicels absent)

Lachenalia splendida *Diels*   (Plate 24d–25a)

splendida: brilliantly coloured flowers

A gregarious, often dwarf *Lachenalia* frequenting sandy flats on the Knersvlakte mainly in the Vanrhynsdorp and Bitterfontein districts, and northwards to Garies. It was previously known as *L. roodeae* Phillips.

The bulb produces two unmarked, lanceolate leaves and the peduncle is usually slightly swollen just below or at the base of the inflorescence. The sessile flowers are oblong-campanulate in shape; the outer perianth segments are pale blue at the base, shading to white or pale lilac, and have greenish-brown gibbosities, while the protruding inner segments are dark lilac with a purple central stripe.

This attractive, very floriferous species is ideally suited to both pot and garden culture.

Flowering period	July–August
Height	60–250 mm
Cultivation performance	Excellent

Lachenalia ventricosa *Schltr.* ex *W.F. Barker*   (Plate 25b)

ventricosa: refers to the swollen peduncle

This often robust *Lachenalia* occurs in the Klawer and Clanwilliam districts, frequenting moist sandy areas, usually at high altitude.

The relatively small bulb produces a single lanceolate or lorate, channelled leaf, which is yellowish-green in colour and usually has an undulate margin. The peduncle is conspicuously swollen just below the inflorescence and this is a very distinctive feature of the species. The inflorescence consists of sessile, oblong-urceolate flowers; the outer perianth segments are very pale blue at the base, shading to dull white and have green or brown gibbosities, while the much longer inner segments are pale yellow, shading to white at the margin, and have a green zone near the tips. The apex of the inflorescence usually consists of numerous sterile flowers.

L. ventricosa is an attractive species but is unfortunately difficult to maintain under cultivation.

Flowering period	August–September
Height	200–480 mm
Cultivation performance	Poor

Subgroup 2b: **Inflorescence subspicate** (pedicels up to 2 mm long)

Lachenalia barkeriana *U.Müller-Doblies et al.*   (Plate 25c)

barkeriana: after Miss W. F. Barker, botanist and former Curator of the Compton Herbarium, Kirstenbosch

A very unusual species which could easily be mistaken for a member of the genus *Polyxena* which it resembles and to which it is closely related, but the latter has equal perianth segments whereas in *Lachenalia* the segments are always at least slightly zygomorphic. It occurs in deep sand in the area between Vanrhynsdorp, Kliprand and Loeriesfontein.

The bulb produces between three to nine erect, linear leaves which are channelled and unmarked. The congested inflorescence is produced at ground level and consists of numerous erect, cylindrical flowers; the perianth segments are greenish or suffused with red and the inner segments protrude. The stamens have very well-exserted filaments which are swollen in the upper half. As with the closely related *L. pusilla,* the peduncle is not visible during the flowering period but elongates during the fruiting stage to facilitate seed dispersal.

L. barkeriana will appeal to the specialist grower as an interesting pot subject.

Flowering period	May–July
Height	15–20 mm
Cultivation performance	Excellent

Lachenalia comptonii *W.F. Barker* (Plate 26a–b)

comptonii: after Prof. R. H. Compton, botanical collector and Director of the National Botanical Gardens until 1953

A heavily scented, late-flowering Karoo species, occurring in the Worcester, Ceres, Laingsburg and Sutherland districts, and as far north as Calvinia. It frequents flat sandy areas and often grows in colonies.

The bulb produces one or two lanceolate, dark green leaves which can be smooth, but which are usually covered with long, simple hairs on the upper surface. It is a proteranthous species in the wild, as its leaves have usually completely withered by the time the flowers open. The inflorescence consists of widely-campanulate, white flowers; the inner and outer perianth segments each have a green keel, while the outer segments have green gibbosities. The well-exserted stamens have conspicuous purple filaments.

L. comptonii requires excellent drainage under cultivation and is highly recommended as a pot subject.

Flowering period	September–October
Height	50–200 mm
Cultivation performance	Good

Lachenalia physocaulos *W.F. Barker* (Plate 26c)

physocaulos: refers to the swollen peduncle and rachis

Occurs in the Robertson and Swellendam districts in sandy ground.

The usually single, linear-conduplicate leaf is glaucous and widens suddenly into a white membranous clasping base. The peduncle and rachis are swollen, most conspicuously below the inflorescence, and are distinctly marked with maroon. The inflorescence is of the subspicate type and bears numerous campanulate flowers; the outer perianth segments are very pale blue at the base, shading to white and have greenish-brown gibbosities, while the protruding inner segments are pale magenta with brownish-green keels.

This attractive species is as yet poorly known in cultivation and appears to be difficult to bring into flower.

Flowering period	August–September
Height	130–300 mm
Cultivation performance	Fair

Subgroup 2c: **Inflorescence subspicate or racemose** (pedicels up to 2 mm long or longer)

Lachenalia gillettii *W.F. Barker* (Plate 26d)

gillettii: after Mr. J. B. Gillett, who discovered this species in 1930

A localized species in the Piketberg and Citrusdal districts which has been rather infrequently collected.

It produces two lorate, bright green leaves which are unmarked, but which have depressed longitudinal veins on the upper surface. The inflorescence consists of oblong-campanulate flowers; the outer perianth segments are white, tinged with lilac at the base, and have green gibbosities, while the protruding inner segments have magenta-coloured tips.

This species will probably have potential as a pot subject when material becomes available.

Flowering period	August–September
Height	120–220 mm
Cultivation performance	Unknown

Lachenalia haarlemensis *Fourc.* (Plate 27a)

haarlemensis: after the town Haarlem, southern Cape

A seldom seen species from the southern Cape, occurring in the Avontuur and Haarlem districts.

One or two erect, linear or lanceolate leaves are produced, which are conspicuously banded with maroon on the clasping base. The inflorescence consists of numerous small, campanulate flowers; the outer perianth segments are pale greenish-grey with slightly darker gibbosities, and the inner segments protrude. The well-exserted stamens have conspicuous mauve filaments.

There are both good and poor forms of this species, and it makes a satisfactory pot subject.

Flowering period	September–October
Height	120–200 mm
Cultivation performance	Excellent

Lachenalia klinghardtiana *Dinter* (Plate 27b)

klinghardtiana: after the Klinghardt mountains, SWA/Namibia

Occurs in the Richtersveld and the south-western corner of SWA/Namibia.

The bulb usually produces a single glaucous, lanceolate leaf which may be plain or lightly spotted on the upper surface. The clasping leaf-base can be banded or spotted with purple, and the peduncle is usually swollen, often conspicuously, just below the inflorescence. The inflorescence consists of dull white, oblong-campanulate flowers; the outer perianth segments each have a conspicuous brown gibbosity, and may or may not be spotted with purple, while the protruding inner segments have a brown marking near the tips.

L. klinghardtiana is short-lived under cultivation, requiring excellent drainage and the constant attention of the specialist grower in order to survive.

Flowering period   July
Height   60–160 mm
Cultivation performance   Poor

Lachenalia latifolia *Tratt.*   (Plate 27c–d)

latifolia: broad-leaved

A late-flowering, very distinctive species from the southern Cape, occurring from Swellendam to George.

It has a rather large bulb, and the two ovate, prostrate leaves may be plain or have irregularly-scattered pustules on the upper surface. Conspicuous longitudinal veins are present on the upper surface, and the species belongs to the small group with proteranthous leaves; in the wild, these have usually dried up completely by the time the flowers open. The inflorescence consists of numerous widely-campanulate flowers, all with recurved segments; both the inner and outer perianth segments are white, each with a dark pink central stripe.

Apart from its attractiveness, this species should also be grown for its distinctively heavy fragrance.

Flowering period   September–November
Height   150–300 mm
Cultivation performance   Good

Lachenalia mathewsii *W.F. Barker*   (Plate 28a)

mathewsii: after Mr J. W. Mathews, first Curator of Kirstenbosch Botanic Garden

Restricted to the Vredenburg district on the Cape west coast, this species was, until recently, considered extinct in nature, not having been re-collected for almost forty years.

The bulb produces two very distinctive glaucous, unmarked leaves which are narrow-lanceolate and taper to a long terete apex. The inflorescence consists of bright yellow, oblong-campanulate flowers; the outer perianth segments each have a conspicuous green gibbosity, while the protruding inner segments have a central green spot near the tips.

Horticulturally, this species is one of the most desirable in the genus, being very floriferous and multiplying rapidly.

Flowering period   September
Height   100–200 mm
Cultivation performance   Excellent

Lachenalia purpureo-caerulea *Jacq.*   (Plate 28b)

purpureo-caerulea: purplish-blue flowers

A late-flowering species with a very restricted distribution in the Darling

Lachenalia moniliformis
(Worcester)

31b. *Lachenalia moniliformis*
(Worcester)

31c. *Lachenalia montana*
(in habitat, Hermanus)

31d. *Lachenalia montana*
(Hermanus)

32a. *Lachenalia multifolia*
(Karoo Poort)

32b. *Lachenalia polyphylla*
(Tulbagh)

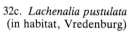

32c. *Lachenalia pustulata*
(in habitat, Vredenburg)

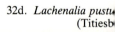

32d. *Lachenalia pustu*
(Titiesb

district. It was probably once a common plant in this area, but due to agricultural activity and encroaching alien vegetation is now considered an endangered species.

The bulb produces two lanceolate or lorate leaves which are proteranthous in the wild and densely pustulate on the upper surface. The inflorescence consists of heavily-scented, widely-campanulate flowers; the outer perianth segments vary from blue or white at the base, shading to magenta or purple and have greenish-brown gibbosities, while the slightly longer inner segments are magenta with darker tips and are distinctly broader.

It is one of the most desirable of the late-flowering lachenalias and deserves to be grown widely.

Flowering period	October–November
Height	100–280 mm
Cultivation performance	Excellent

Lachenalia pusilla *Jacq.* (Plate 28c–d)

pusilla: plants are very small

This is one of the most unusual members of the genus as it flowers at ground level and produces a rosette of prostrate leaves. It is closely related to *L. barkeriana*, and has previously been included in the genus *Polyxena* as *P. pusilla* (Jacq.) Schltr. *L. pusilla* has a fairly wide distribution; from Nieuwoudtville south to the Cedarberg and Piketberg, to Vredenburg on the west coast, inland to the Worcester district, and south to Bredasdorp and Swellendam. It favours sandy habitats and is one of the earliest flowering species.

The bulb produces between four and six leaves which are usually spotted with green or brown above, and which can vary from linear to lanceolate. The peduncle is not visible during the flowering period, but in the fruiting stage it elongates considerably to facilitate seed dispersal. The inflorescence is a congested raceme of white, erect cylindrical flowers, with conspicuously exserted stamens. Flowers have a strong, unique scent.

L. pusilla is an interesting pot subject for the specialist grower.

Flowering period	April–June
Height	10–40 mm
Cultivation performance	Excellent

Lachenalia salteri *W.F. Barker* (Plate 29a)

salteri: after Capt. T. M. Salter, botanical collector and author

A late-flowering, coastal species occurring mainly between Betty's Bay and Bredasdorp, but is also recorded from the Cape Peninsula. Its habitat is often marshy areas around seasonal pools, and it is a variable species as regards overall plant size and flower colour.

The bulb produces two lanceolate, leathery leaves which may be plain or have large brown blotches, mainly on the upper surface. The inflorescence consists of numerous oblong-campanulate flowers. Flower colour ranges from cream to reddish-purple, but most typically it is pale blue at the base of the outer perianth segments, shading to white with brownish-purple gibbosities, while the protruding inner segments vary in shades of pink.

L. salteri flowers at a time when most other species have long since finished and it is a rewarding, highly desirable subject under cultivation.

Flowering period | October–December
Height | 150–350 mm
Cultivation performance | Excellent

Subgroup 2d: **Inflorescence racemose** (pedicels usually longer than 2 mm)

Lachenalia anguinea *Sweet*   (Plate 29b)

anguinea: snake-like markings on leaves

The long, flaccid, banded leaf of this *Lachenalia*, seen in the veld amongst rocks could well be momentarily mistaken for a snake, and it is a coastal species occurring in red sand from the Richtersveld as far south as the Piketberg district.

The bulb produces a single lanceolate or narrow-lanceolate leaf which is conspicuously banded with green on the lower surface and with maroon on the clasping base. Plants often become robust, and the usually sturdy peduncle bears a raceme of numerous campanulate, pendulous, cream-coloured flowers produced on very long pedicels. The outer perianth segments have green gibbosities and the inner segments protrude slightly.

A most attractive species with a singular, strong scent. It requires excellent drainage under cultivation.

Flowering period | July–September
Height | 100–350 mm
Cultivation performance | Good

Lachenalia duncanii *W.F. Barker*   (Plate 29c)

duncanii: after G. D. Duncan, who discovered this species in 1985

At present only known to occur near Kliprand in south-western Bushmanland.

It produces two glaucous, lanceolate-falcate leaves which are channelled above and which may be plain or have purplish blotches on the upper surfaces. The leaves are leathery, undulate, and usually distinctly crisped along their margins. The inflorescence consists of numerous oblong-campanulate, cream-coloured flowers; the outer perianth segments have green keels shading to darker green gibbosities and the protruding inner segments have pale green keels shading to darker green at the tips. The filaments of the exserted stamens are magenta in the upper third.

An interesting species for the specialist grower.

Flowering period	August–September
Height	150–180 mm
Cultivation performance	Good

Lachenalia esterhuysenae *W.F. Barker* (Plate 29d)

esterhuysenàe: after Miss Elsie Esterhuysen, renowned botanical collector of plants from the Cape mountains

A seldom seen, montane species from the upper reaches of the Cedarberg.

The bulb produces one or two linear, terete, blue-green leaves which are unmarked and do not clasp the base of the very slender peduncle. The inflorescence consists of cream-coloured, campanulate flowers; the outer perianth segments have pale green gibbosities and the slightly longer inner segments have a green marking near the tips. The flowers are borne on long, spreading pedicels.

This dainty, late-flowering *Lachenalia* would probably only appeal to the specialist grower and is at present unknown in cultivation.

Flowering period	October–December
Height	150–450 mm
Cultivation performance	Unknown

Lachenalia glaucophylla *W.F. Barker*

glaucophylla: grey-green leaves

A small-flowered, usually dwarf species, known mainly from the Calvinia district, but also recorded much further north from the Kamiesberg.

The bulb produces a distinctive, usually single, unmarked leaf; it is conspicuously curved, narrow-lanceolate, conduplicate and grey-green in colour. The inflorescence consists of very small, widely-campanulate flowers; the outer perianth segments are cream-coloured with green gibbosities, while the slightly longer inner segments are cream-coloured with green keels.

The species is poorly known in cultivation, but makes an attractive pot subject.

Flowering period	October
Height	90–250 mm
Cultivation performance	Good

Lachenalia juncifolia *Bak.*

juncifolia: narrow, grass-like leaves

A variable, often dwarf species, occurring in the Worcester and Ceres districts, south towards Caledon and along the coast to Still Bay. It is frequently associated with limestone outcrops. Two varieties are recognized.

i) var. **juncifolia** (Plate 30a)

The bulb produces two leaves which vary from filiform to linear and can be terete or semi-terete, and usually have conspicuous maroon bands or markings on the leaf-bases. The inflorescence consists of oblong-campanulate flowers which vary considerably in colour from locality to locality; the outer perianth segments are usually white, tinged with pale pink and can have purple, dark pink or green gibbosities. In other forms, the segments are dark pink, or blue at the base shading to pink. The inner segments protrude slightly and have light to dark pink keels.

This variety multiplies rapidly and is recommended for both pot and garden culture.

Flowering period   August–October
Height   70–230 mm
Cultivation performance   Excellent

ii)   var. **campanulata** *W.F. Barker*   (Plate 30b)
campanulata: bell-shaped flowers

This variety appears to be restricted to the coastal belt from Caledon to Riversdale.

The two semi-terete leaves are succulent with a very narrow channel above, and have bases banded with maroon. The inflorescence consists of white, campanulate flowers; the outer perianth segments have deep rose gibbosities, and the slightly longer inner segments have deep rose keels. The stamens are usually less exserted than in var. *juncifolia.*

A most attractive, very floriferous variety, recommended for both pot and garden culture.

Flowering period   September–October
Height   80–300 mm
Cultivation performance   Excellent

Lachenalia latimerae *W.F. Barker*

latimerae: after Dr M. Courtenay-Latimer, former Director of the East London
          Museum

This species is known mainly from the Patensie district in the eastern Cape, but is also recorded from Oudtshoorn, indicating a potentially wider distribution.

Its one or two linear-lanceolate, flaccid leaves are channelled above, and can be plain or spotted with magenta on the clasping base. The inflorescence consists of numerous pale pink or lilac campanulate flowers; the outer perianth segments have greenish-brown gibbosities and the protruding inner segments have a pinkish-brown spot at the tips.

This species is as yet unknown in cultivation, but will have potential as a pot subject when material becomes available.

Flowering period   July–August
Height   150–280 mm
Cultivation performance   Unknown

Lachenalia macgregoriorum *W.F. Barker*   (Plate 30c–d)

macgregoriorum: after the Macgregor family of Glenlyon Farm, Nieuwoudtville, who have played an influential role in nature conservation in the north-western Cape

This rare, late-flowering species is at present only known from the Nieuwoudtville district, where it has been collected on just one occasion.

It belongs to the small group of species with proteranthous leaves; in the wild
the plant comes into flower after the foliage has withered. The bulb produces
two glaucous, linear leaves which are channelled above, and minutely spotted or
banded with maroon in the lower regions. The inflorescence consists of widely
campanulate flowers; the outer perianth segments are dull white with a broad
central maroon stripe, while the inner segments are maroon, slightly longer, and
distinctly broader.

A fairly attractive, unusual species which is ideally suited to pot culture.

Flowering period   October–November
Height   160–360 mm
Cultivation performance   Excellent

Lachenalia moniliformis *W.F. Barker*   (Plate 31a–b)

moniliformis: leaves have a beaded appearance

This dainty species is at present known only from its type locality in the
Worcester district.

Its bulb is characterized in producing numerous stolons from the base, each
terminating in a bulbil produced at ground level. The many grass-like terete
leaves are very distinctive; they have circular, raised fleshy bands along the
upper two thirds of their length, resembling strings of beads. The lower portions
of the leaves are banded with maroon and shade to magenta at the base. The
inflorescence consists of small campanulate flowers; the outer perianth segments
are very pale blue at the base, shading to very pale pink, and have reddish-brown
gibbosities, while the very slightly longer inner segments are white with a
reddish-brown marking near the tips.

A highly desirable pot subject which multiplies rapidly and is very floriferous.

Flowering period   September
Height   120–170 mm
Cultivation performance   Excellent

Lachenalia montana *Schltr.* ex *W.F. Barker*   (Plate 31c–d)

montana: from mountainous terrain

This late-flowering species frequents montane fynbos habitat in the Hermanus, Caledon and Franschhoek districts. As with *L. sargeantii*, it requires fire
to be stimulated into successful flowering in the wild, and may subsequently be
seen in profusion, often in association with *L. peersii*.

The often deeply lodged bulbs produce two conduplicate, linear leaves which are unmarked and loosely clasp the base of the peduncle. The dense inflorescence consists of numerous pendulous, campanulate flowers borne on magenta-coloured pedicels. The cream-coloured outer perianth segments have large green or greenish-brown gibbosities, and the slightly protruding inner segments have a green or greenish-brown marking near the tips.

A very desirable species, but a shy flowerer under cultivation.

Flowering period	October–December
Height	100–330 mm
Cultivation performance	Fair

Lachenalia multifolia *W.F. Barker*   (Plate 32a)

multifolia: many-leaved

A strongly scented species from the Worcester, Ceres and Calvinia districts which is always associated with rocky habitat.

It belongs to the small group with numerous, grass-like leaves; they are terete, yellowish-green in colour and have smooth, white swollen bases. The inflorescence consists of cream-coloured, widely campanulate flowers; the outer perianth segments have pale green gibbosities, while the equal, or very slightly longer, inner segments have a green central marking near the tips.

L. multifolia would no doubt have potential as a container plant, but is as yet unknown in cultivation.

Flowering period	September–October
Height	70–200 mm
Cultivation performance	Unknown

Lachenalia nordenstamii *W.F. Barker*

nordenstamii: after Dr B. Nordenstam, Swedish botanist whose published works include South African genera

A rare, dwarf species from the northern Richtersveld and southern SWA/Namibia, favouring sheltered rock cracks.

The bulb is covered with layers of fibrous tunics, forming a distinct neck, and produces a single glaucous, lanceolate leaf which is unmarked on the upper surface, but banded with maroon on the clasping base and lower surface. The few-flowered inflorescence consists of small, widely campanulate flowers; the perianth segments are brownish, with the inner segments very slightly longer, and all with a central maroon stripe. The well-exserted stamens have comparatively thick, maroon filaments.

An interesting species for the specialist grower, but requires excellent drainage.

Flowering period	June–July
Height	50–120 mm
Cultivation performance	Fair

Lachenalia polyphylla *Bak.*   (Plate 32b)
polyphylla: many-leaved

A rare, very seldom seen species recorded from the Malmesbury, Piketberg and Tulbagh districts, occurring on flat, open ground.

As the specific epithet indicates, it belongs to the small group which have numerous grass-like leaves, and its leaf bases are swollen, maroon-coloured in the upper half and covered with tiny hairs. The inflorescence is a delicate raceme of small, narrowly-campanulate flowers; the outer perianth segments are pale blue at their bases and shade to rose-pink, while the very slightly protruding inner segments are white with a rose-pink central zone and have recurved tips.

L. polyphylla is as yet unknown in cultivation, but not being particularly attractive, it would probably only appeal to the specialist grower.

Flowering period   September–October
Height   60–180 mm
Cultivation performance   Unknown

Lachenalia polypodantha *Schltr.* ex *W.F. Barker*
polypodantha: refers to the many conspicuous, feet-like anthers

A dwarf species with a very restricted distribution in the Springbok district and which has been collected on just two occasions.

It produces a single ovate leaf which is densely covered with stellate hairs on the upper surface. The very short peduncle bears a racemose inflorescence of widely-campanulate, white flowers borne on long pedicels; the outer perianth segments have pale green gibbosities, and the protruding inner segments have a pale green spot near the tips. The well-exserted stamens have conspicuous purple filaments.

The species is as yet unknown in cultivation, but will no doubt have potential as a pot subject when material becomes available.

Flowering period   August–September
Height   50–150 mm
Cultivation performance   Unknown

Lachenalia pustulata *Jacq.*   (Plate 32c–33a)
pustulata: leaves with blisters on upper surface

Often found growing in large colonies, this scented *Lachenalia* has several colour forms, and occurs on flats and rocky outcrops in the Saldanha, Malmesbury, Paarl and Worcester districts, as well as on the Cape Peninsula.

The one or two lanceolate or lorate leaves may or may not be covered with pustules on the upper surface, depending on locality. The inflorescence consists of numerous oblong-campanulate flowers; in the typical colour forms, the outer perianth segments are cream or very pale yellow, and have green gibbosities, while the colour forms found on the west coast in the Saldanha–Vredenburg districts occur in various shades of pink or blue, and have brownish-pink gibbosities. The protruding inner segments have a dark pink or brown, or pale green central marking at the tips. It is very closely related to *L. unicolor*.

The species is very floriferous and highly recommended for both pot and garden culture.

Flowering period	August–October
Height	150–350 mm
Cultivation performance	Excellent

Lachenalia stayneri *W.F. Barker*   (Plate 33b–c)

stayneri: after Mr Frank Stayner, Curator of the Karoo National Botanic Garden from 1959–1969

A very distinctive species occurring in the Worcester and Robertson districts.

It is easily recognized by its two prostrate, lanceolate or lorate leaves which have large, irregularly scattered pustules on the upper surface. The inflorescence consists of numerous campanulate flowers; the outer perianth segments are very pale blue at the base, shading to cream and have reddish-brown gibbosities, while the slightly longer inner segments are cream with a reddish-brown spot near the tips.

This species multiplies very slowly under cultivation, but is nevertheless a recommended pot subject.

Flowering period	August–September
Height	120–300 mm
Cultivation performance	Excellent

Lachenalia unicolor *Jacq.*   (Plate 33d–34c)

unicolor: single-coloured flowers

A very variable, complex species with a wide distribution; from Nieuwoudtville south to Riebeek-Kasteel, inland as far as the Worcester district and south to Somerset West. It is absent from the Cape Peninsula.

The two lanceolate or lorate leaves are usually densely pustulate on the upper surface and vary in colour from pale green to dark maroon, depending on locality. The inflorescence consists of numerous oblong-campanulate flowers, and the outer perianth segments vary in colour from cream with pale blue bases and green gibbosities, to various shades of lilac, pink, magenta, blue or purple with darker gibbosities. The inner segments protrude and the stamens are conspicuously exserted. Certain forms are pleasantly scented. The species is very closely related to *L. pustulata* and may well have to be included with it at some future stage.

L. unicolor is floriferous, multiplies rapidly and is horticulturally one of the most rewarding species.

Flowering period	September–October
Height	80–300 mm
Cultivation performance	Excellent

33a. *Lachenalia pustulata*
(Darling)

33b. *Lachenalia stayneri*
(Worcester)

33c. *Lachenalia stayneri*
(Worcester)

33d. *Lachenalia unicolor*
(Vanrhynsdorp)

34a. *Lachenalia unicolor*
(Citrusdal)

34b. *Lachenalia unicolor*
(Porterville)

34c. *Lachenalia unicolor*
(Tulbagh)

34d. *Lachenalia violacea* var. *viola*
(Klaw

35a. *Lachenalia violacea* var. *violacea*
(Nigramoep)

35b. *Lachenalia violacea* var. *violacea*
(Sutherland)

35c. *Lachenalia violacea* var. *glauca*
(in habitat, Kamieskroon)

35d. *Lachenalia whitehillensis*
(Whitehill)

36a. *Lachenalia whitehillensis*
(Whitehill)

36b. *Lachenalia zebrina* forma *zebrina*
(Calvinia)

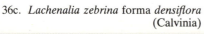

36c. *Lachenalia zebrina* forma *densiflora*
(Calvinia)

36d. *Lachenalia zebrina* forma *densiflo*
(Calvin

Lachenalia violacea *Jacq.*

violacea: violet-coloured flowers

An extremely variable, often robust species, its distribution extending from the southern Richtersveld throughout Namaqualand, the Knersvlakte and as far south as the Clanwilliam district. Two varieties are recognized.

i)   var. **violacea**   (Plate 34d–35b)

This is the most widespread and variable of the two varieties. The bulb produces one or two lanceolate leaves which vary greatly in colour and markings from locality to locality; in some forms they are bright green, unmarked and not undulate, whereas the form occurring near Bitterfontein, for example, has a single glaucous leaf which is banded on the clasping base, spotted above and below, and has a conspicuously undulate, crisped margin. The peduncle is often swollen just below the inflorescence, a distinctive feature of the latter being its long pedicels. The raceme consists of numerous campanulate flowers. The outer perianth segments vary in colour from pale blue-green or white at the base, shading to pale magenta or purple, and the gibbosities are brown. The very slightly protruding inner segments vary in shades of purple or violet.

This variety is highly recommended as a pot subject.

Flowering period	July–September
Height	100–350 mm
Cultivation performance	Good

ii)   var. **glauca** *W.F. Barker*   (Plate 35c)

glauca: pale greyish-magenta flowers

This variety occurs in the Springbok, Kamieskroon and Clanwilliam districts.

The bulb produces a single lanceolate, plain green leaf with an undulating margin. The inflorescence consists of numerous oblong-campanulate flowers; the outer perianth segments are grey-blue at the base, shading to pale magenta with slightly darker gibbosities, and the very slightly protruding inner segments are also pale magenta, giving the whole inflorescence a general glaucous appearance. This variety is also characterised in having a distinctive cocoanut fragrance.

L. violacea var. *glauca* deserves attention as a pot subject but requires excellent drainage.

Flowering period	September
Height	100–230 mm
Cultivation performance	Fair

Lachenalia whitehillensis *W.F. Barker*   (Plate 35d–36a)

whitehillensis: after Whitehill Station, near Matjiesfontein

A delicate species occurring mainly in the Matjiesfontein district, but also recorded further north from Sutherland, indicating a potentially wider distribution.

The bulb produces a single, very distinctive leaf; it is narrow-lanceolate or linear, conduplicate and conspicuously banded with maroon on the lower surface, shading to magenta toward the clasping base. The peduncle and rachis are also conspicuously spotted. The inflorescence consists of campanulate, scented flowers; the outer perianth segments are pale blue at the base shading to cream and have reddish-brown gibbosities, while the protruding inner segments are cream with a reddish-brown spot near the apex and a pale blue central stripe.

A most attractive, but short-lived species under cultivation, requiring excellent drainage.

Flowering period   September
Height   150–360 mm
Cultivation performance   Poor

Lachenalia zebrina *W.F. Barker*

zebrina: refers to the resemblance of the striped leaves to a zebra's hind leg

A very variable and often robust species with a wide distribution in the Karoo and Knersvlakte; from Matjiesfontein to Vanrhynsdorp, common in the Calvinia district and as far east as Carnarvon. It used to be included within *L. anguinea*, but is now regarded as distinct due to its different seed pattern and distribution. Two named forms are recognized within the species.

i)   forma **zebrina**   (Plate 36b)

This is the most typical form and produces a single falcate or lanceolate leaf which is conduplicate or channelled above, and distinctly banded with maroon on the lower surface and clasping base. The upper leaf surface is glaucous, often with undulate margins. The inflorescence is a lax raceme of small campanulate flowers, borne on long pedicels; the outer perianth segments are cream-coloured with a brown or green tinge, and the inner segments protrude slightly. The stamens protrude conspicuously.

ii)   forma **densiflora** *W.F. Barker*   (Plate 36c–d)

densiflora: flowers are numerous and congested

This form occurs within the distribution range of forma *zebrina*, but is less common.

It differs from forma *zebrina* in its denser, narrower inflorescence and its flowers are borne on comparatively shorter pedicels.

Both forms are highly recommended for container cultivation, but require excellent drainage.

Flowering period   August–October
Height   150–300 mm
Cultivation performance   Good

GLOSSARY

Apex	The tip
Arillode	A false aril, an appendage of the seed
Campanulate	Bell-shaped
Cartilaginous	Tough and hard
Conduplicate	Folded together lengthwise
Crisped	Curled
Declinate	Bent or curved downwards or forwards
Falcate	Curved like a sickle
Filiform	Thread-like
Gibbosity	A swelling
Glaucous	Blue-green, grey-green, grey-blue or greyish-magenta
Inflorescence	The flower cluster
Keel	A ridge
Lanceolate	Lance-shaped; much longer than broad, widening above the base and tapering to a pointed end
Linear	Long and narrow
Locule	A compartment of the ovary
Lorate	Strap-shaped; with a blunt end
Ovate	Shaped like a longitudinal section of an egg; the basal portion broader than the upper portion
Pedicel	The stalk of a flower
Peduncle	The stalk of the inflorescence
Pendulous	Drooping: hanging downwards
Perianth	The two floral envelopes considered together
Prostrate	Lying flat on the ground
Proteranthous	Leaves mature before the flowers
Racemose	Simple, elongated inflorescences with flowers borne on distinct pedicels
Rachis	The axis bearing the flowers
Segment	Part of the perianth
Sessile	Without a pedicel
Spicate	Simple, elongated inflorescences with sessile flowers
Stellate	Star-like
Subspicate	Simple, elongated inflorescences with flowers borne on very short pedicels
Terete	Circular in transverse section
Tunicate	Having surrounding coats or layers
Undulate	Wavy up and down
Urceolate	Urn-shaped
Ventricose	Swollen or inflated
Zygomorphic	Divisible into equal halves in one plane only

BIBLIOGRAPHY

BAKER, J. G., 1897. *Lachenalia* Jacq. In: Thistleton–Dyer, W. T. (ed.), *Flora Capensis* 6: 421–436. Reeve & Co., London.

BARKER, W. F., 1930. *Lachenalia. Jl Bot. Soc. S. Afr.* XVI: 10–13.

BARKER, W. F., 1933. *Lachenalia elegans. Flow. Pl. S. Afr.* 13 t. 508.

BARKER, W. F., 1933. *Lachenalia gillettii. Flow. Pl. S. Afr.* 13 t. 506.

BARKER, W. F., 1950. *Lachenalia* Jacq.f. In: Adamson & Salter (eds.), *Flora of the Cape Peninsula*: 198–202. Juta, Cape Town.

BARKER, W. F., 1966. The rediscovery of two South African plants and the renaming of another. *Bot. Notiser* 119: 201–207.

BARKER, W. F., 1969. A new combination in *Lachenalia* with notes on the species. *Jl S. Afr. Bot.* 35 (5): 321–322.

BARKER, W. F., 1972. A new species of *Lachenalia* from the south-western Cape. *Jl S. Afr. Bot.* 38 (3): 179–183.

BARKER, W. F., 1978. Ten new species of *Lachenalia* (Liliaceae). *Jl S. Afr. Bot.* 44 (4): 391–418.

BARKER, W. F., 1979. Ten more new species of *Lachenalia* (Liliaceae). *Jl S. Afr. Bot.* 45 (2): 193–219.

BARKER, W. F., 1980. *Lachenalia trichophylla. Flow. Pl. Afr.* 46 t. 1808.

BARKER, W. F., 1983. A list of the *Lachenalia* species included in Rudolph Schlechter's collections made in 1891–1898 on his collecting trips in Southern Africa, with identifications added. *Jl S. Afr. Bot.* 49 (1): 45–55.

BARKER, W. F., 1983. Six more new species of *Lachenalia* (Liliaceae). *Jl S. Afr. Bot.* 49 (4): 423–444.

BARKER, W. F., 1984. Three more new species of *Lachenalia* and one new variety of an early species (Liliaceae). *Jl S. Afr. Bot.* 50 (4): 535–547.

BARKER, W. F., 1987. Five more new species of *Lachenalia* (Liliaceae–Hyacinthoideae) —four from the Cape Province and one from southern South West Africa/Namibia. *S. Afr. J. Bot.* 53 (2): 166–172.

BAYER, M. B., 1982. *The new Haworthia Handbook.* National Botanic Gardens of South Africa, Cape Town.

CROSBY, T. S., 1986. The Genus *Lachenalia. The Plantsman* Vol. 8 (3): 129–166.

DELPIERRE, G. R., & DU PLESSIS, N. M., 1973. *The winter-growing Gladioli of South Africa.* Tafelberg, Cape Town.

DE WET, J. M. J., 1957. Chromosome numbers in the Scilleae. *Cytologia* 22: 145–159.

DUNCAN, G. D., 1986. The re-discovery of *Lachenalia mathewsii* W. Barker. *Veld & Flora* 72 (2): 40–41.

DUNCAN, G. D., 1987. *Lachenalia macgregoriorum. Flow. Pl. Afr.* 49 t. 1951.

INGRAM, J., 1966. Notes on the cultivated Liliaceae, 4. *Lachenalia. Baileya* Vol. 14 (3): 123–132.

MOFFETT, A. H., 1936. The cytology of *Lachenalia. Cytologia* 7: 490–498.

MÜLLER-DOBLIES, U., NORDENSTAM, B. and MÜLLER-DOBLIES, D., 1987. A second species in *Lachenalia* subgen. *Brachyscypha* (Hyacinthaceae): *Lachenalia barkeriana* sp. nov. from southern Little Namaqualand. *S. Afr. J. Bot.* 53 (6): 481–488.

NEL, D. D., 1983. Rapid propagation of *Lachenalia* hybrids *in vitro. S. Afr. J. Bot.* 2 (3): 245–246.

ORNDUFF, R., and WATTERS, P. J., 1978. Chromosome numbers in *Lachenalia. Jl S. Afr. Bot.* 44 (4): 387–390.

REID, C., 1985. *Lachenalia aloides. Flow. Pl. Afr.* 48 t. 1910.

SALTER, T. M., LEWIS, G. J. and BARKER, W. F., 1949. *Plantae Novae Africanae. Jl S. Afr. Bot.* XV: 37–39.

SMITH, C. A., 1966. *Common names of South African Plants. Lachenalia* Jacq. p. 541. Dept. Agricultural Technical Services.

STEARN, W. T., 1983. *Botanical Latin.* David & Charles, London.

WAGER, V. A., 1984. *Plant Pests and Diseases.* Jonathan Ball, Johannesburg.

INDEX

The names of species upheld in this work are in bold type; synonyms are in italics.